NOT A GOOD SIGN

NOT A GOOD SIGN

DON HOUSE

With a Foreword by Larry Foley

PHOENIX INTERNATIONAL, INC.
FAYETTEVILLE

Designed by John Coghlan
Edited by Debbie Self

Inquiries should be addressed to:
Phoenix International, Inc.
17762 Summer Rain Road
Fayetteville, Arkansas 72701
Phone (479) 521-2204
www.phoenixbase.com

Library of Congress Cataloging-in-Publication Data

House, Don, 1951–
  Not a good sign / Don House ; with a foreword by Larry Foley.
    p. cm.
  ISBN 978-0-9824295-0-1 (alk. paper)
  1. Arkansas—Social life and customs—Pictorial works. 2. Scenic byways—Arkansas—Pictorial works. 3. Signs and
signboards—Arkansas—Pictorial works. 4. Errors—Arkansas—Pictorial works. 5. Arkansas—History, Local—Pictorial
works. 6. Middle West—Social life and customs—Pictorial works. 7. Scenic byways—Middle West—Pictorial works.
8. Signs and signboards—Middle West—Pictorial works. 9. Black-and-white photography—Arkansas. 10. Black-and-
white photography—Middle West. I. Title.
  F412.H68 2010
  976.7—dc22
                                                                                              2009035261

To Denise, who has always been willing to take the road less traveled, and navigate to boot

FOREWORD

S*ign, sign, everywhere a sign… Slow down, you're movin' too fast…*

Those are lyrics from a couple of songs popular back in the early 1970s when I was a teenager living in Tennessee, and later Texas, when we took long road trips back to Arkansas to visit my grandparents in Van Buren. It is amazing what impacts us, and what we remember, if we were ever paying attention in the first place. We tend to recall most, visual images that conjure up some kind of emotion, whether it is humor, amazement, or some sort of offbeat irony.

An earlier generation will recall the iconic Burma-Shave billboards that once dotted the landscape, when family vacations were always taken in big Detroit-made sedans, along narrow two-lane roads that never bypassed a small town or big city along the way. And there were always signs, of every sort, telling us to "See Rock City" or letting us know how far to the "Dog 'n' Suds," where the root beer floats were world famous.

A college friend and I took a rambling road trip 30 plus years ago into the Great Smokey Mountains. We noticed that families would get out of their cars and march the tribe next to the big sign along the road for the photographic backdrop, forgetting about the scenic panoramic vista they drove to see. For whatever reason, they needed a "sign" to make sure everybody would remember their destination. Whenever possible, they'd do their best to also try and work their old Ford, Chevy, or Dodge in the picture. The station wagon, billboard, and family in the frame at the same time—now that was photographic brilliance!

When I was starting my career, I often drove by a little family-owned washateria in North Little Rock, Arkansas, with a homemade sign out front advertising, "Drop Of Service." I miss that sign, and those days, but not as much as one might think. You see, those signs are still out there, if you slow down, get off the interstate, and take the time to look around at an America that still has its regional curiosities and local flavor. Get a road map—the real folding kind you can hold in your hands—and ramp off the four-lane. That's where you see the really unusual stuff.

I've spent my professional days as a documentary filmmaker, looking for interesting people and places that seem to have escaped general attention. I find it almost sacrilegious to drive by a historical marker without pulling over and reading "what happened on this spot" back in the 1800s.

Through the lens of his camera, and his own curiosity, Don House shows us what's out there for us all to see—what is really going on in our country. These are snapshots in present-day history, frozen in time for our private contemplation. And when you have finished enjoying Don's work, go and load up the children or grandkids in search of your own signs.

As Simon and Garfunkel sang to us way back there on the road somewhere, "slow down" and "make the morning last." It might even make you feel groovy.

Larry Foley
Fayetteville, Arkansas
July 30, 2009

INTRODUCTION

When we get these thruways across the whole country, as we will and must, it will be possible to drive from New York to California without seeing a single thing.

—John Steinbeck, *Travels with Charley*, 1962

*W*ho's *my Daddy?* The large brown eyes plead with passersby. The child clutches a teddy bear. Thousands of people are driving by each hour, yet no one stops to help. In fact, it is against the law. Emergencies only, and stopping to stare at a billboard does not qualify, but I am mystified by the statistical analysis it implies, the question it begs—how does a Houston-based clinic, nine hundred miles south, determine that this spot in rural Missouri is the perfect place to advertise DNA testing services?

I am breaking my own rule of avoiding interstates. I-44—the freeway that cuts diagonally through Missouri from Saint Louis, crosses the border, and ends up in Oklahoma City—has a new rest area. A considerable amount of time and effort have been spent in its making, and forgive me, but I've come to mock it. The designers have put together a tribute to Route 66, the legendary east-west highway from Chicago to Los Angeles, perhaps the most romanticized American roadway of the twentieth century. And what was chosen to demonstrate the uniqueness and romance of what once was? Why, reproductions of old signs, of course. Signs for diners, tourist attractions, gas stations, and motels; inside and outside the visitor center, painted on picnic shelters, planted by the front door, and hovering over the vending machines. Even out of context, stripped of all the voices and laughter, sights and sounds, they have the power of nostalgia for several generations of Americans.

Timing is everything. Perhaps if I had been born a few years earlier, or a few years later, I would look at the interstate freeway system in a completely different way. It might conjure words like progress, national security, jobs, commerce, and convenience, instead of scandal, corruption, racism, waste, boondoggle, environmental disaster, and personal tragedy. But the impressions of childhood are difficult to exorcise. Coming of age when the interstate system was being seriously questioned by a growing number of Americans, I have maintained that skepticism into adulthood.

If the families gunning their engines to merge back on the interstate are aware of the irony in which they have been participants—that they are reentering the very raison d'extinction of all that was good about Route 66, there are no outward signs, and if they are considering letting their children experience a little of the magic about which they were just lectured, by detouring at the next exit and taking the back roads to Saint Louis and beyond, they don't mention it to me. Thus, the whole facility takes on the appearance of the worst sort of historical marker, the kind that essentially states: *Something really cool and wonderful used to be here before we destroyed it. Just wanted to let you know. Have a nice day.*

What Route 66 embodies in a concentrated form, intensified by hindsight, is the fundamental difference between the interstate system and almost every other road in the nation. They all were meant to *connect* communities while the interstate was designed to *bypass* them. From my home in northwest Arkansas, it is a two-hundred-mile drive to the capitol. I can grab I-540 south to I-40 and take it straight east to Little Rock, or I can head east on highway 16, catch highway 23 south to highway 10, then east. The difference between the two routes is wrapped around the dictum that we all know by rote, *that it is the journey rather than the destination that should have our attention.* During my alternate route east, I pass through over twenty-five communities, drive over the highest point in the state, eat lunch in Paris, stick my bare feet in the cool waters of four rivers, climb to the top of Pinnacle Mountain, and see signs, hundreds of signs.

At the end of a long drive, there is fatigue, of course, regardless of route, but one is the fatigue of boredom and repetition, and one is the fatigue of experience and exercise. Every traveler is aware of the inverse relationship between speed and connection—walk and interact with the world the most, fly and interact the least—and we look for some balance, some sweet compromise, which for me turns out to be back roads, windows down, 45mph, with frequent stops. And those stops are most often to photograph signs.

Back road signs speak a local dialect, not a national tongue. They are the voice of the community, calling to us who pass, telling us where to stop, how to park, what to eat, what to buy, what to see, what's important. I read them all, and when circumstances are right, when human error, or line of sight, or word phrasing add an unintended slant, I grab my camera. But I am laughing with them, not at them. No mistake has been made that I wouldn't make, no wording that I couldn't have chosen, no misspelling that I haven't made myself. The signs are a conversation between the people who live there and me, the visitor. We recognize what we have in common, our shared flawed humanity, and it makes me feel at home, somehow welcome. Without talking to anyone, I feel like I know them, and perhaps that is the essence of it—choosing the back road is an expression of love.

Of course, as with any love affair, there can be moments of regret, a questioning of choices made, prices paid, time wasted, the road not taken, perhaps. Then, you are cruising slowly down highway 74 near the small community of Tuttle, Arkansas. You pass a cemetery, and glancing at the stones and names, you suddenly swerve off the road and stop, not out of respect for the dead, but because there is a building right next to the cemetery, a building with a large sign that seems to float over the tombstones: *Dwight's Body Shop*. At that moment, your lover has just turned to you and smiled and blown you a kiss, and life is good.

I leave you with an interesting historical note. In 1961, a Missouri engineer by the name of Rex Whitton became the federal highway administrator. He has been credited with rescuing the interstate system from scandal, funding problems, and public apathy that in the early sixties were fueling calls by Congress and the press to end it. In an interview with a trade magazine after he retired, he described how he and his wife enjoyed traveling, but they avoided the freeways he had helped to build, preferring— "driving on the little back roads, keeping a map of each one we travel."[1]

Don House
July 4, 2009

---

1. Used with the kind permission of the Federal Highway Administration, from an article by Richard F. Weingroff.

❖ Fat hogs for sale

Highway 7, just south of Jasper, Arkansas, 2008

What is it about this combination of words? Does anyone ever picture a lean hog? But adding fat conjures up images of cuteness. A hog, you buy to eat, a fat hog you buy as a pet. And the placement on this stretch of highway, with no farm in sight, makes it seem like a trap. A well-armed pig waiting in the underbrush for a car to stop.

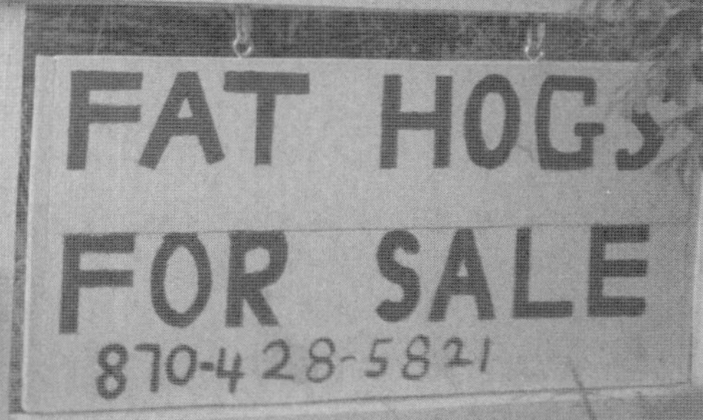

FAT HOGS
FOR SALE
870-428-5821

2   ❖   Welcome to the City of Clermont

Highway 50, Clermont, Florida, 2003

Vermin? Alligators? Amoeba? What parent, after reading this sign, could possibly lay out a blanket, squirt on some sun block, and watch the kids head for the water?

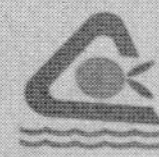

Welcome to the City of Clermont
Waterfront Park

The beach and park are open from sunrise to sunset.
No Lifeguards on duty. Swim at your own risk.
Swim in designated areas only.
WARNING: Deep holes beyond designated swim area.
No Jumping or diving off of piers.
Children 12 and under are required to be accompanied by an adult.
Keep the beach and park clean by placing trash in garbage containers.
The following are prohibited: Alcohol, Drugs, Controlled Substances,
Glass Containers, Pets, Fires and Camping.
WARNING: Amoeba, Alligators, Snakes, Vermin and other
Native Wildlife may be present.
STATE LAW: Do not feed Alligators.
Report nuisance animals and violators to (352) 394-5588

La playa y el parque estan abierto de la salida hasta la puesta del sol.
No hay salvavidas en servicio. Nade a su propio riesgo.
Nader en las areas designada solamente.
ADVERTENCIA: Areas profundas despues de areas designadas.
No saltar de los muelles.
Ninos menores de 12 anos deberan estar acompanados por un adulto.
Mantenga el area limpia deposite la basura en el zafacon.
Se prohibe traer bebidas alcolicas, sustancias controladas,
envases de cristal, mascotas, fogatas o acampar.
ADVERTENCIA: Animales tales como Amebas, lagartos, culebras y otros
animales nativos pueden encountrarse en el parque.
LEY ESTATAL: No alimente los lagartos.
Reporte violadores del reglamento al (352) 394-5588

❖ Free farm

Highway 65, across from Cummins prison, near Grady, Arkansas, 2003

Location, location, location. Apparently, when you own property adjoining a prison, marketing opportunities are limited.

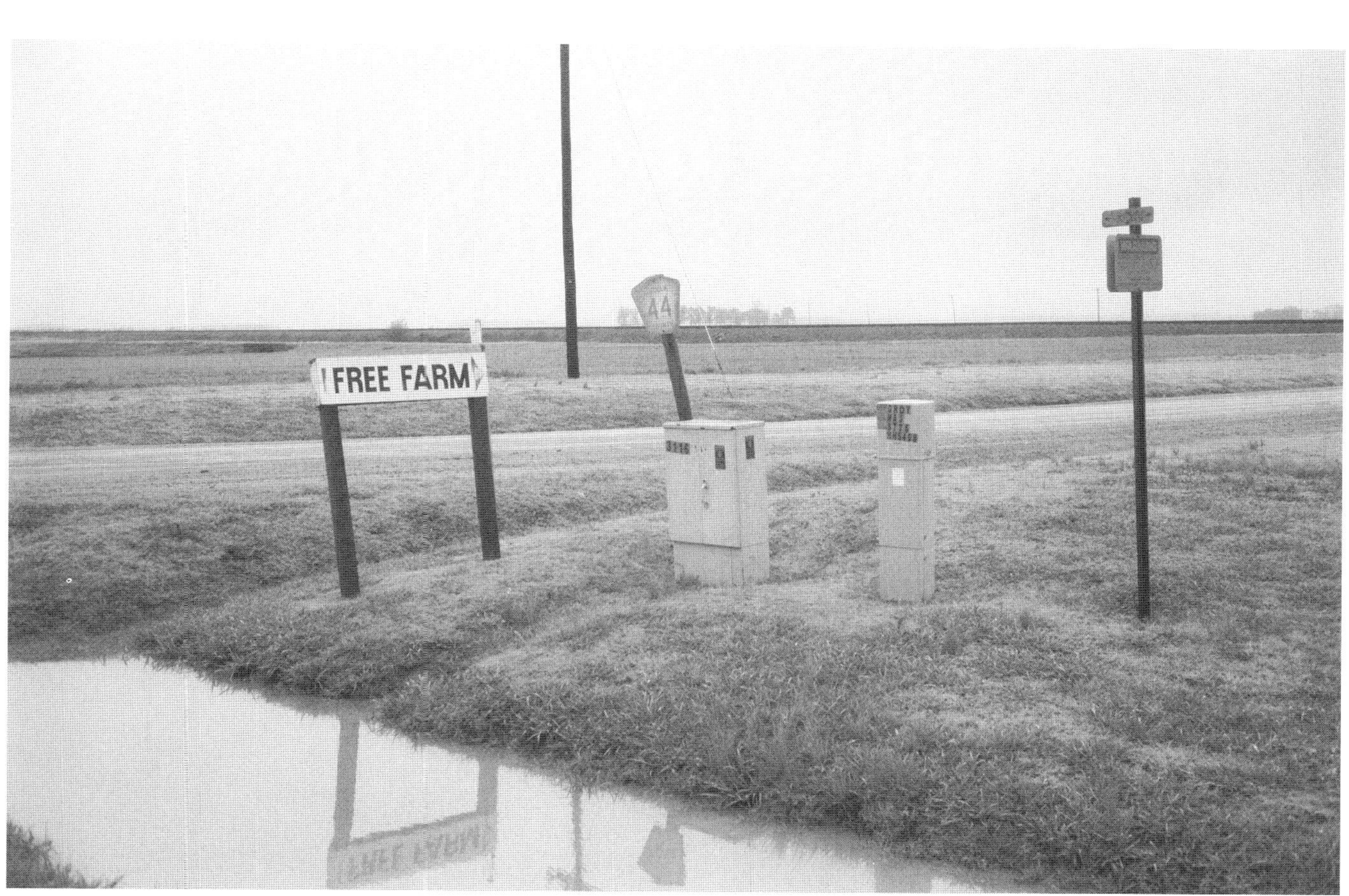
FREE FARM
44

❖   Lake ahead

Highway 12, east of Rogers, Arkansas, 2008

Stopping to photograph this sign in the floodwaters of Beaver Lake, I was unprepared to wade, having just come from a client visit, dressed in street clothes and shoes. But, if I've learned one thing over the years, it's not to put off a photograph when the getting's good, so I just walked out through the dark water. People were sitting on their porches nearby, and I figured they were about to be entertained by the sight of a crazy man with a very expensive camera disappearing beneath the waves, but I made the shot and returned safely. As I stepped out of the water, a woman was standing in her yard looking at me. She yelled, "Mister, you ain't got no waders on!" I just waved, poured the water out of my shoes, acted like it was an everyday thing, and drove off.

LAKE
AHEAD
NO FIRES
NO CAMPING
CAMPING PERMITTED
ONLY IN

5   ❖   No vending allowed

Highway 261, Moki Dugway, Utah, 2007

The question, really, is to which side of the fence this warning applies, and how, what, and why one might possibly vend.

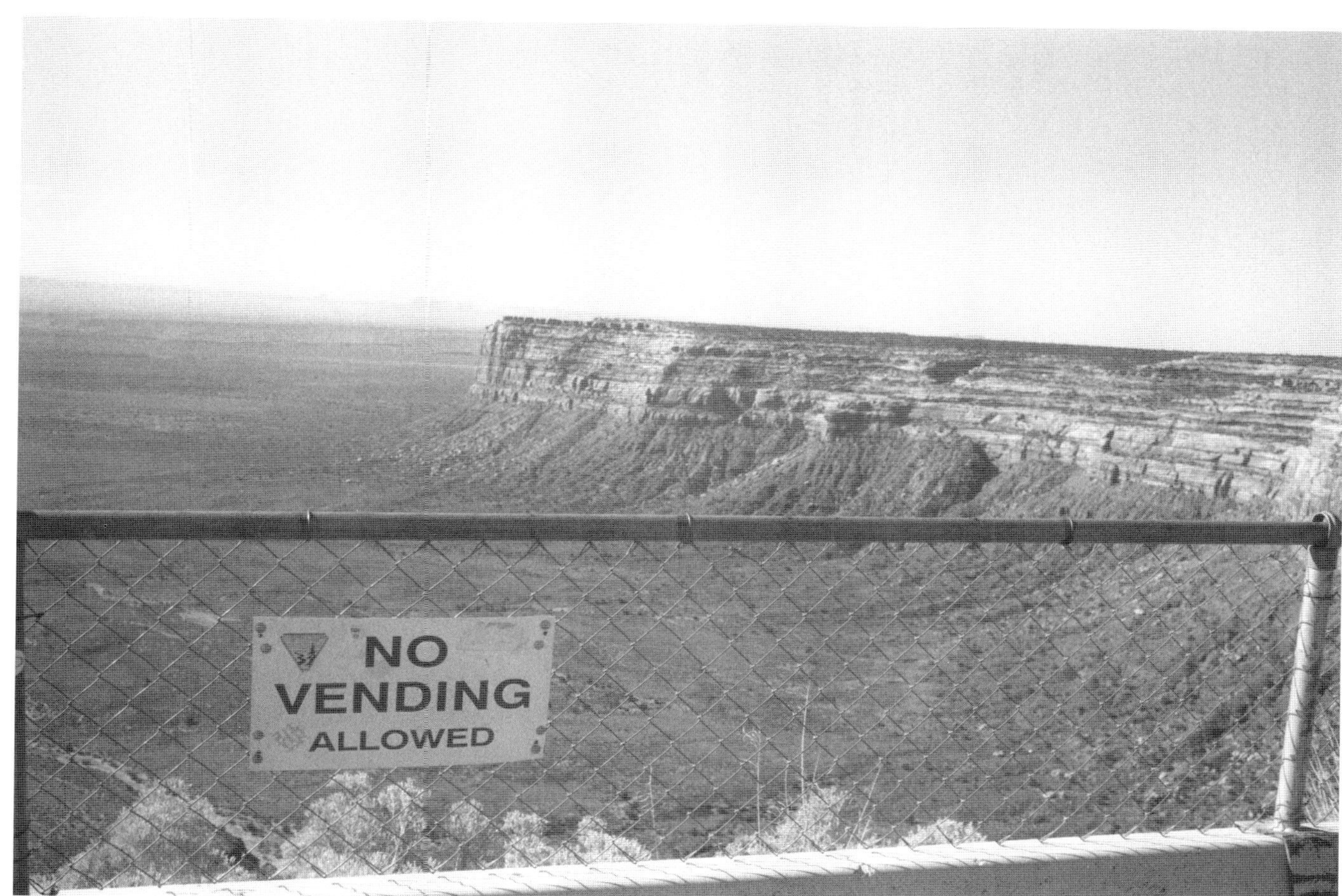

NO
VENDING
ALLOWED

6   ❖   Spiritual fruit

There had been a flurry of press reports dealing with the controversial ordainment of an openly gay priest in the Episcopal church just before this sign appeared. My first reaction was surprise at what appeared to be a show of support for the beleaguered minister, and from a denomination not usually associated with social liberalism. I have come to doubt that interpretation.

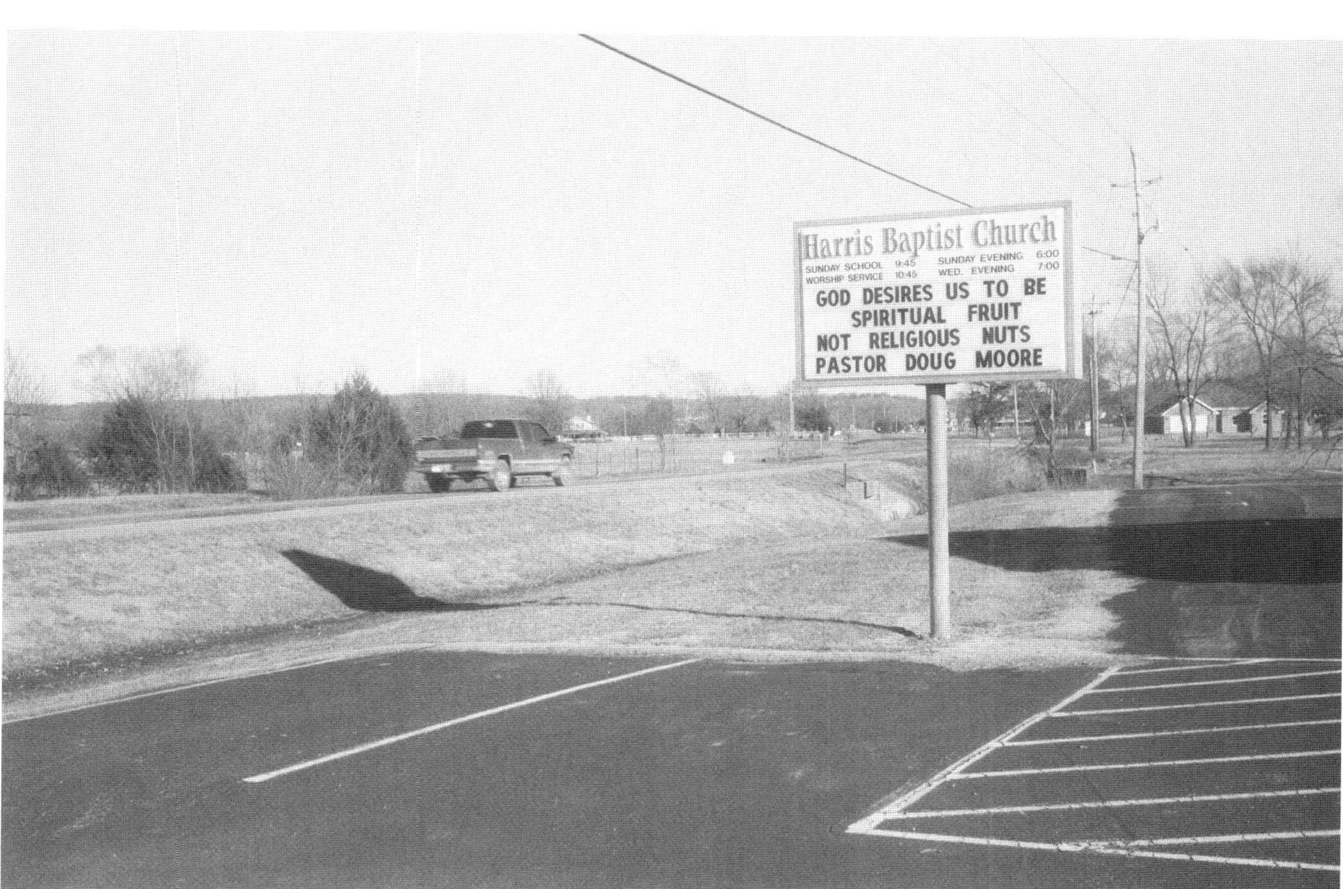

Harris Baptist Church
SUNDAY SCHOOL   9:45   SUNDAY EVENING   6:00
WORSHIP SERVICE   10:45   WED. EVENING   7:00
GOD DESIRES US TO BE
SPIRITUAL FRUIT
NOT RELIGIOUS NUTS
PASTOR DOUG MOORE

7. ❖   Unsafe when under water

Highway 220, near Cedarville, Arkansas, 2000

An example of a mystery sign. Like the one on a bathroom wall at a gas station in Conway, Arkansas: DO NOT URINATE ON THE FLOOR! Wouldn't the audience have to fall into one of two categories?

People who would never even consider urinating on the floor and don't need a sign to remind them.

People who *would* urinate on the floor, and trust me, will not be deterred by a sign.

Is there the third group that the sign implies? People who would stop and look at the lake before them, read the sign, and think, "Hmmmm, I was planning on driving underwater, but if you say it's unsafe, then I guess I won't."

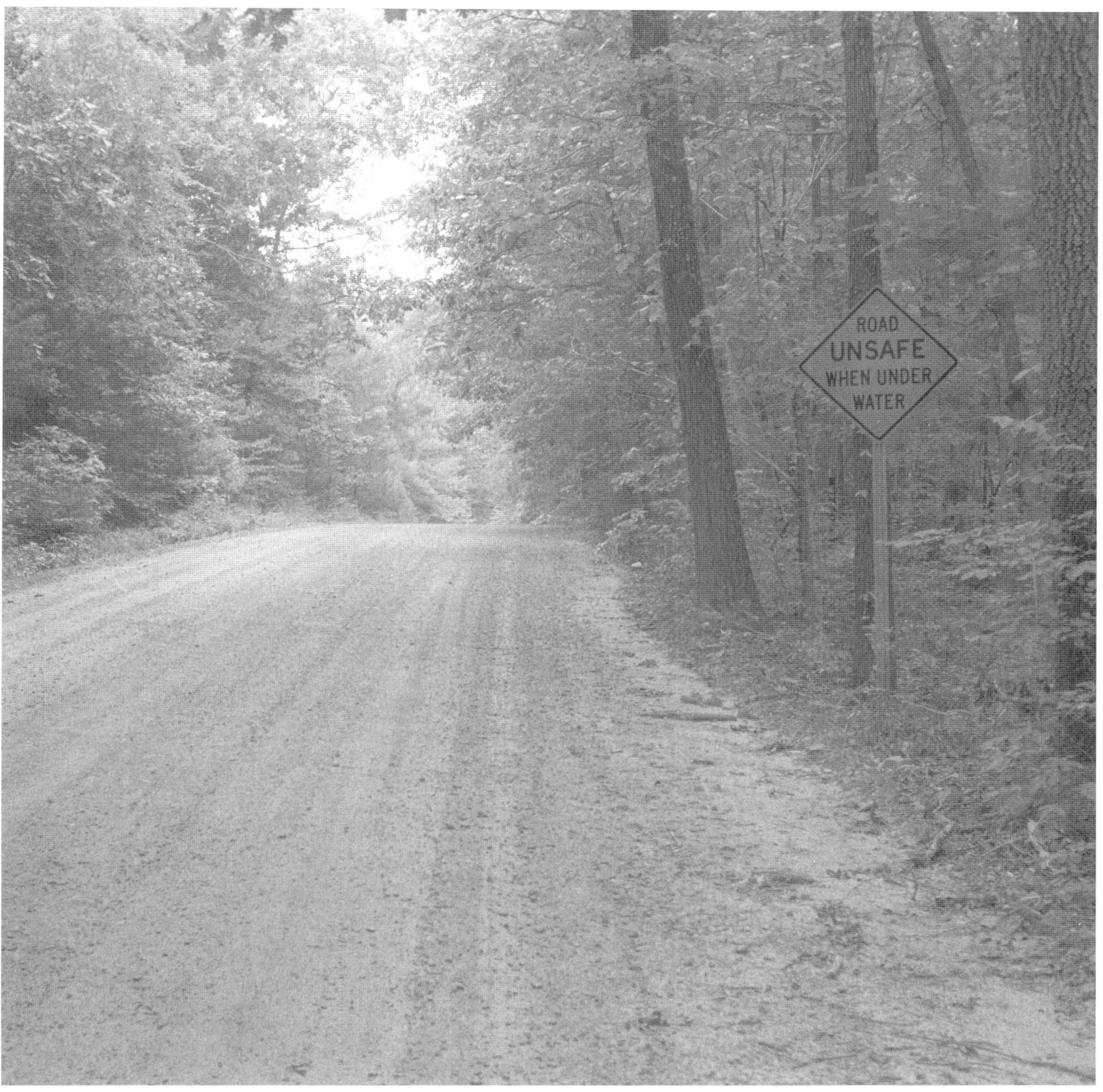
ROAD
UNSAFE
WHEN UNDER
WATER

8 ❖ All you can eat all day

Old Highway 412, Locust Grove, Oklahoma, 2007

I used to frequent a small Salvadoran cafe in Springdale, Arkansas. The owner complained to me one day that he'd been told by his customers that he would never make it if he didn't offer an all-you-can-eat buffet—a mandatory American thing. He resisted because he felt the quality of the food would suffer, and, sure enough, he went out of business. I was thinking of him when I saw this Mexican restaurant in a small Oklahoma town. The owner's final frustrated capitulation to gluttony.

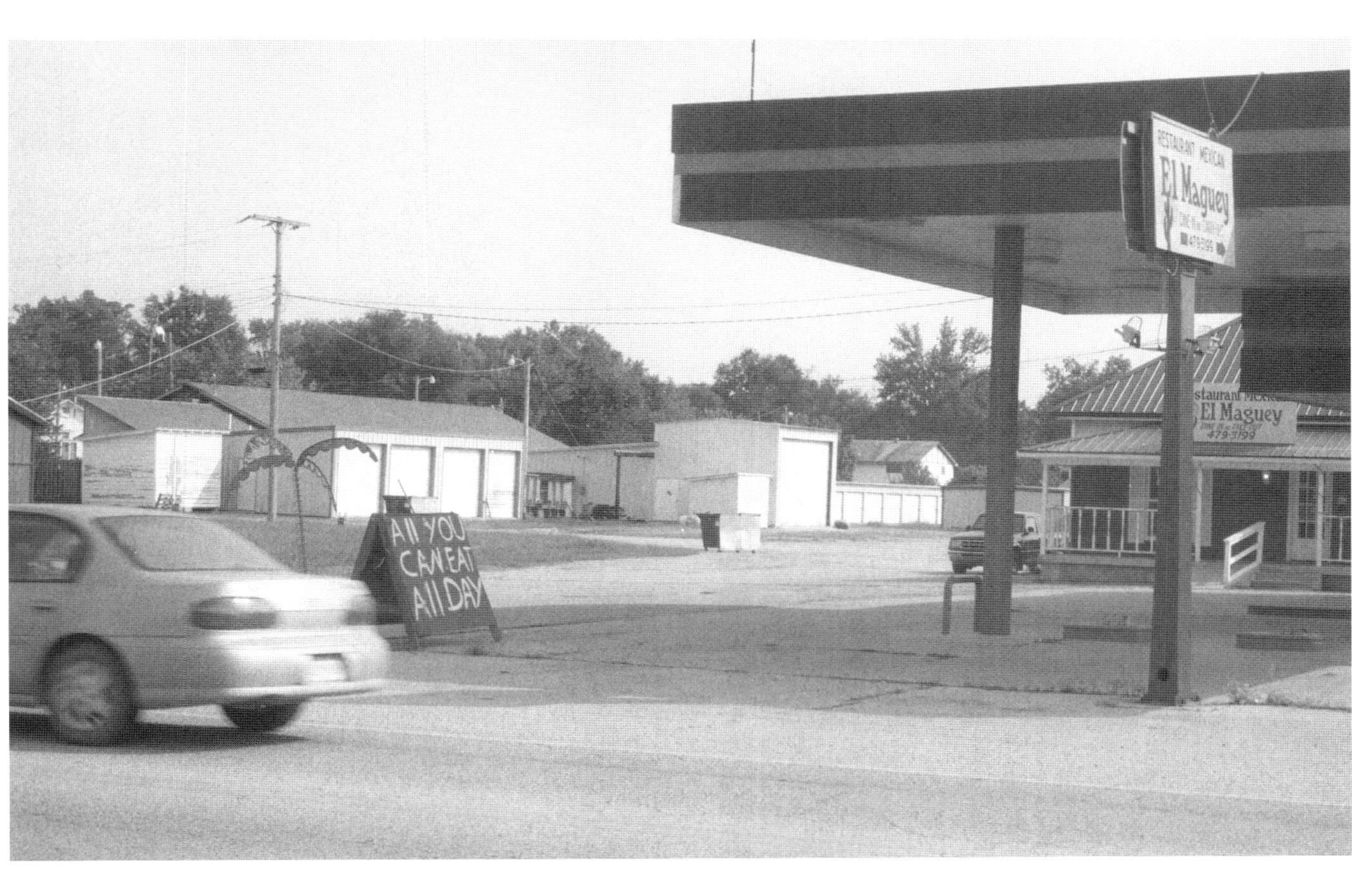

RESTAURANT MEXICAN
El Maguey
All YOU
CAN EAT
All DAY

9

❖   Access area regs

Highway 59, Illinois River, near Tahlequah, Oklahoma, 1997

But have a nice day! At least they didn't prohibit photography.

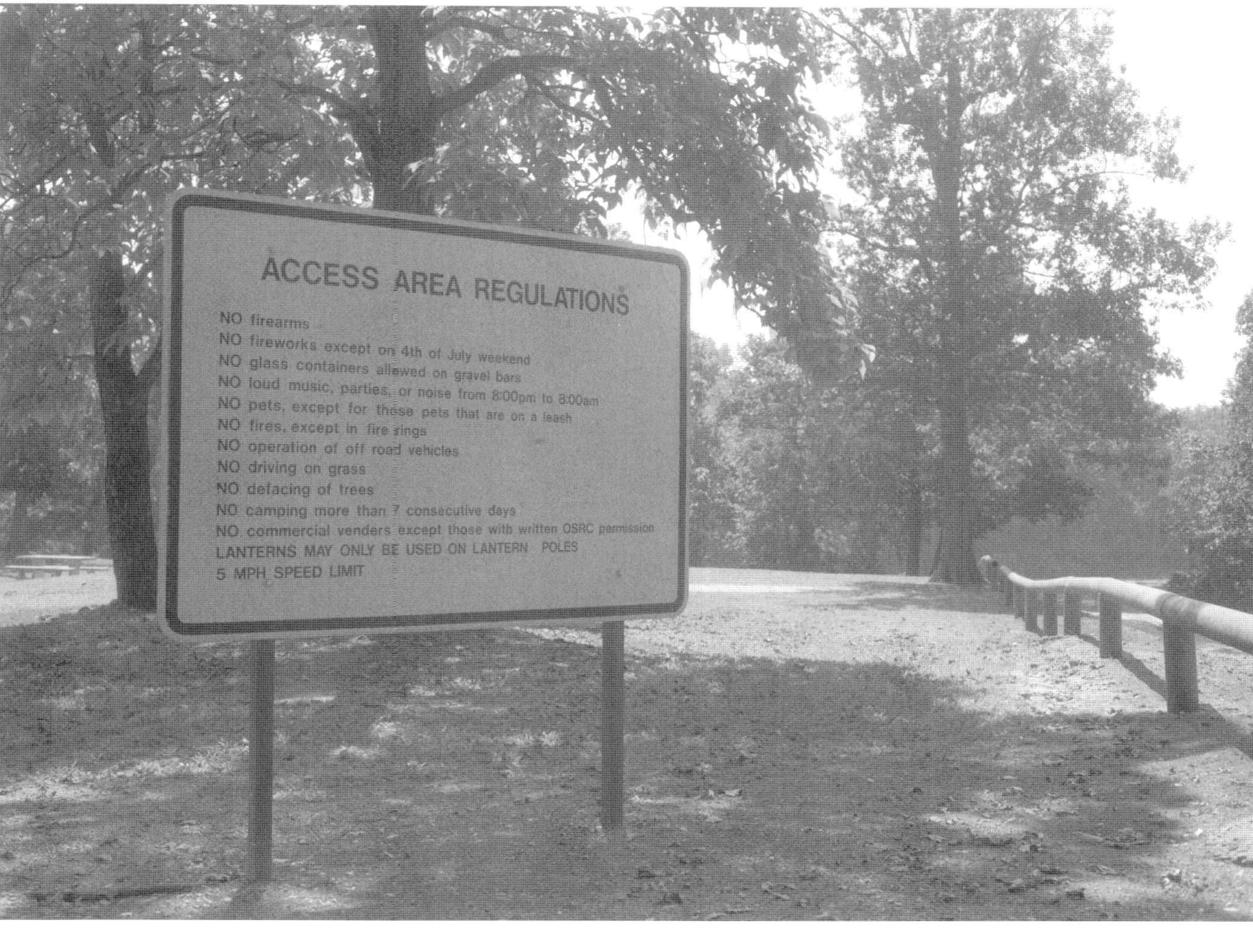

ACCESS AREA REGULATIONS
NO firearms
NO fireworks except on 4th of July weekend
NO glass containers allowed on gravel bars
NO loud music, parties, or noise from 8:00pm to 8:00am
NO pets, except for those pets that are on a leash
NO fires, except in fire rings
NO operation of off road vehicles
NO driving on grass
NO defacing of trees
NO camping more than 7 consecutive days
NO commercial venders except those with written OSRC permission
LANTERNS MAY ONLY BE USED ON LANTERN POLES
5 MPH SPEED LIMIT

❖   America

Highway 37, just south of Goreville, Illinois, 2003

In 2003, thousands of people who drove this route had the same thought, "So *here's* where you've been hiding."

AMERICA

❖   Broken arm rock

Highway 23, near White Rock Recreation Area, Arkansas, 2006

So, what I want to know is whether the person who wrote this was wearing a cast a week after the event, and wanting to warn others, or (and I hope this is true) did she crawl back to her vehicle on the day of the fall, get a can of paint, and refuse to leave before completing an act of defiant and angry defacement.

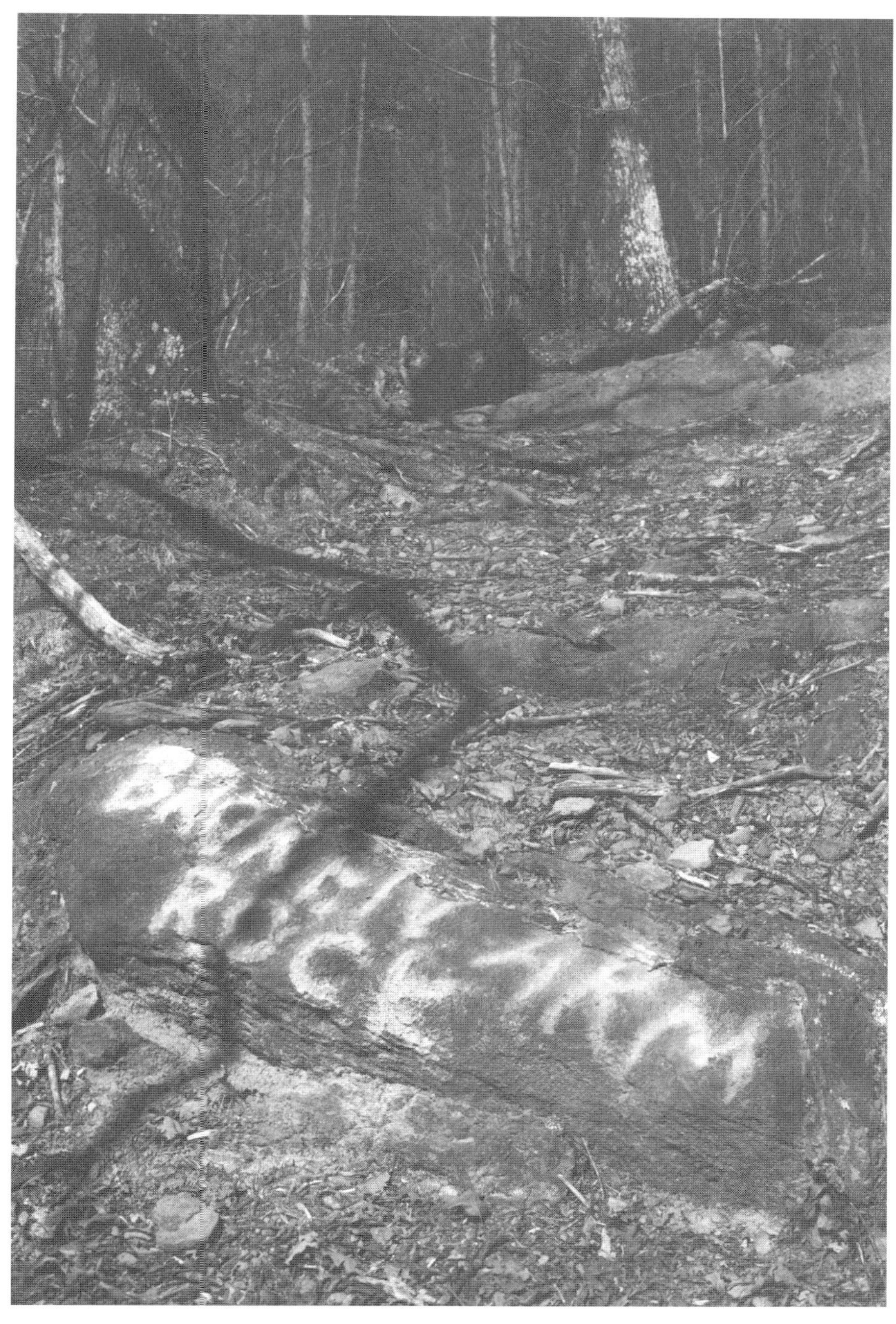
BRONEY
RIDGE FARM

12 ❧ Hair and photography

Highway 37, Benton, Illinois, 2001

Photography's a tough way to make a living. In my family, there's a standing joke that when my sign changes from *Fine Art Photography* to *Don's Frames and Fries*, they'll know I'm in trouble.

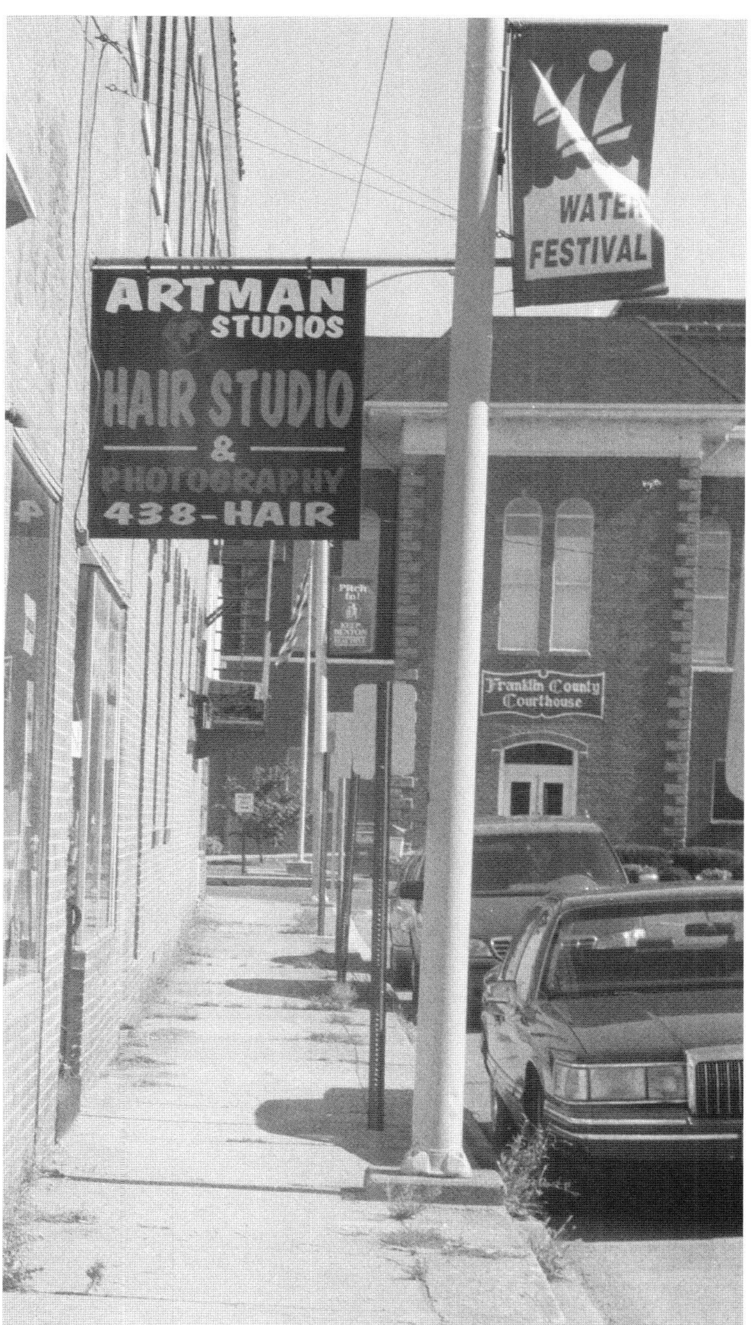
WATER
FESTIVAL
ARTMAN
STUDIOS
HAIR STUDIO
&
PHOTOGRAPHY
438-HAIR
Franklin County
Courthouse

❖ Consultants, Inc.

Highway 112, near Tontitown, Arkansas, 2004

Well, as your consultants, the first thing we advise you to do is get rid of that damned tree.

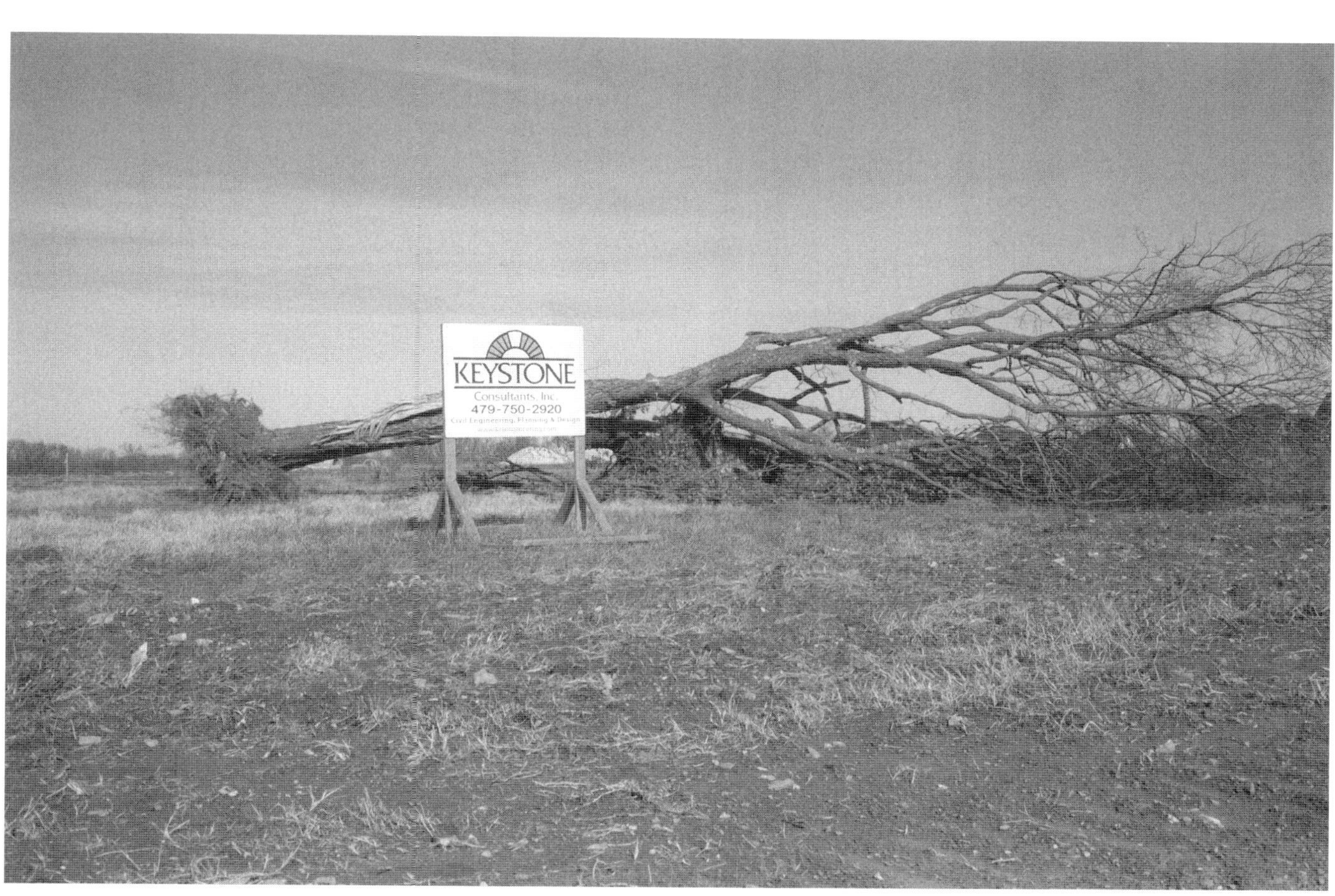
KEYSTONE
Consultants, Inc.
479-750-2920
Civil Engineering, Planning & Design
www.keystoneeng.com

Highway 65, Ferriday, Louisiana, 2001

On assignment for *American Forests* magazine photographing champion trees in Arkansas and Louisiana—the kind of paid job that only comes along once in a blue moon—I had spent a week in the presence of giants, living out of my truck, blissfully unconnected to current events. Passing this intersection, I wondered if a reference was being made to the mental cleansing that libraries offer, or whether I'd better turn on the news.

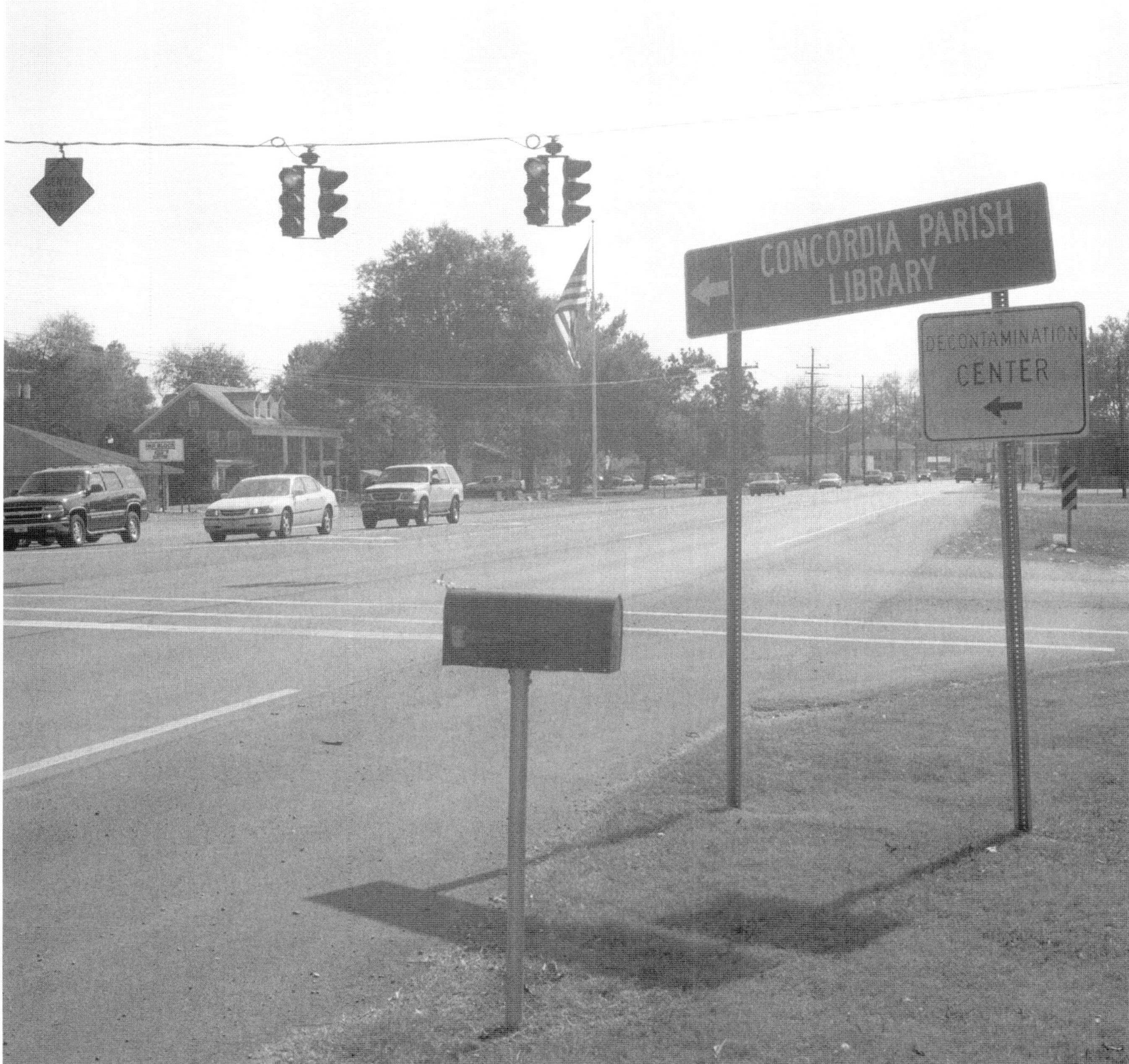
CONCORDIA PARISH
LIBRARY
DECONTAMINATION
CENTER

15 ❖ Duck pluckers

Highway 165, near Stuttgart, Arkansas, 2007

To anyone familiar with the enormous phenomenon of duck hunting in southeastern Arkansas, the business and religion and tradition of it, this sign makes perfect sense. To someone passing through, it can be an enigma, like the one seen farther south in pecan country that boldly announced: "BUY, SELL, CRACK."

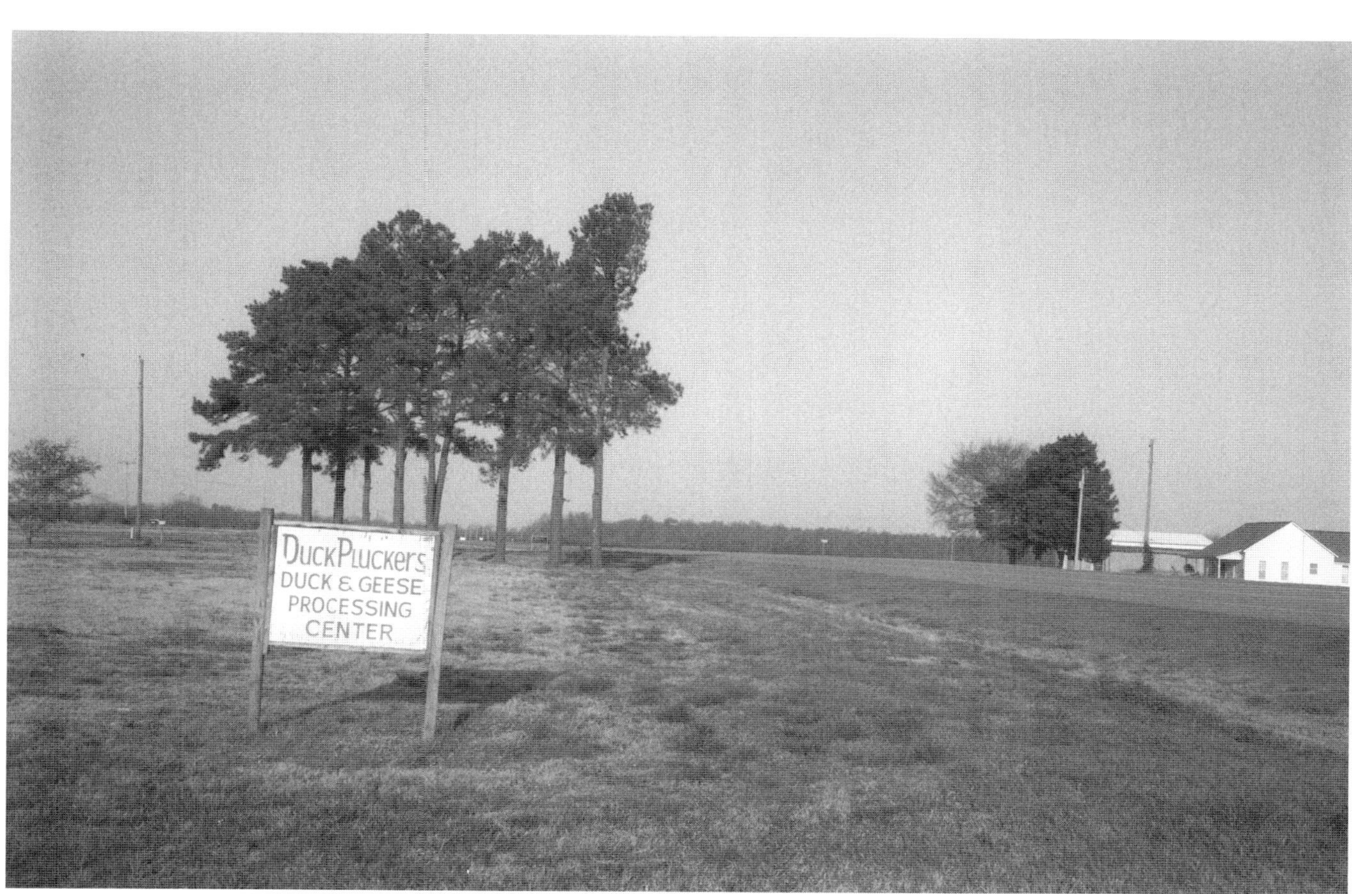
DuckPluckers
DUCK & GEESE
PROCESSING
CENTER

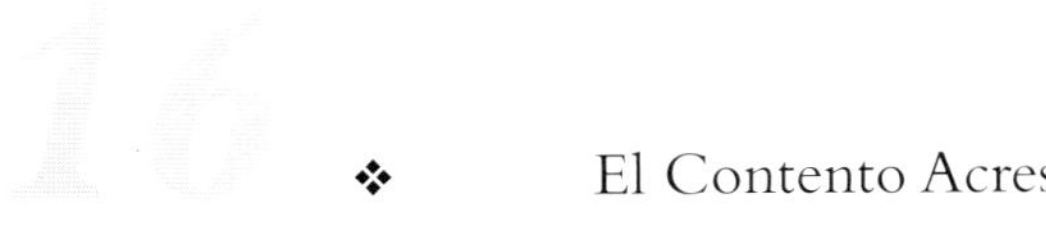

❖   El Contento Acres

Highway 72, near Bentonville, Arkansas, 2009

Northwest Arkansas became a boom area a few years ago. One of the changes it brought was exclusive communities with creative names. Perhaps someone simply got transferred and had to leave, but driving by on a beautiful Ozark spring morning, it felt sadder, as if contentment had not been found behind those walls.

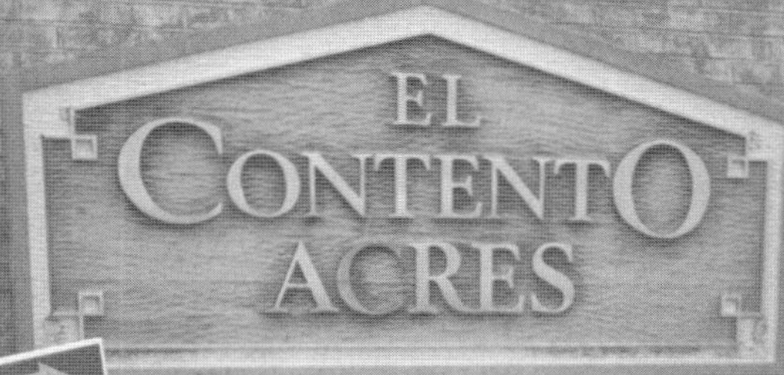

EL
CONTENTO
ACRES
FOR SALE
www.NickyDeu.com

17 ❖ Heave

Highway 8, Cherokee, Oklahoma, 2001

Thanks, but I think I'll pass.

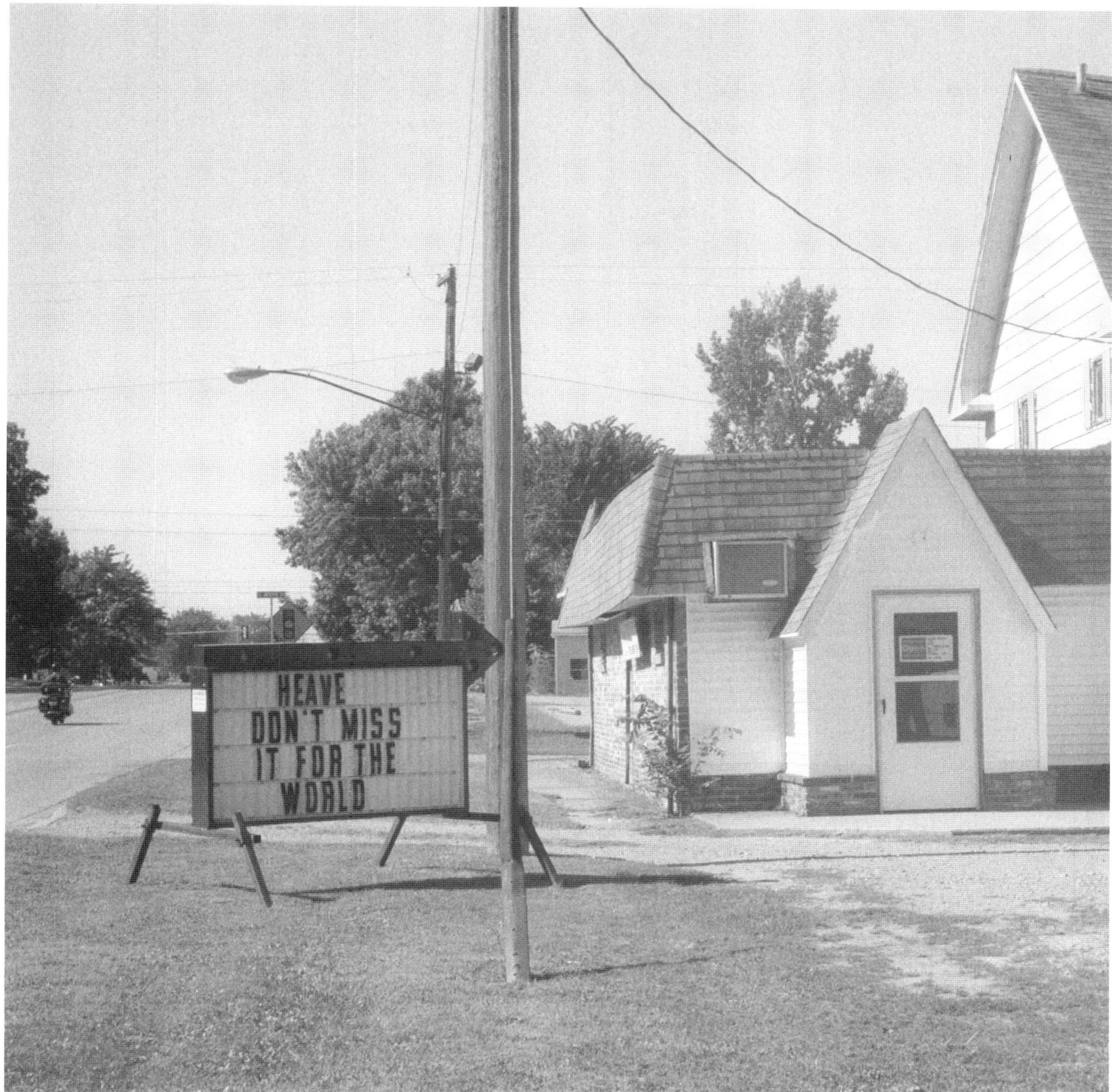
HEAVE
DON'T MISS
IT FOR THE
WORLD

Highway 64, Farmington, New Mexico, 2007

Who knows if this has been an effective campaign, but it sure worked on me. I was feeling guilty just parking *next* to the video store, His eyes following my every move.

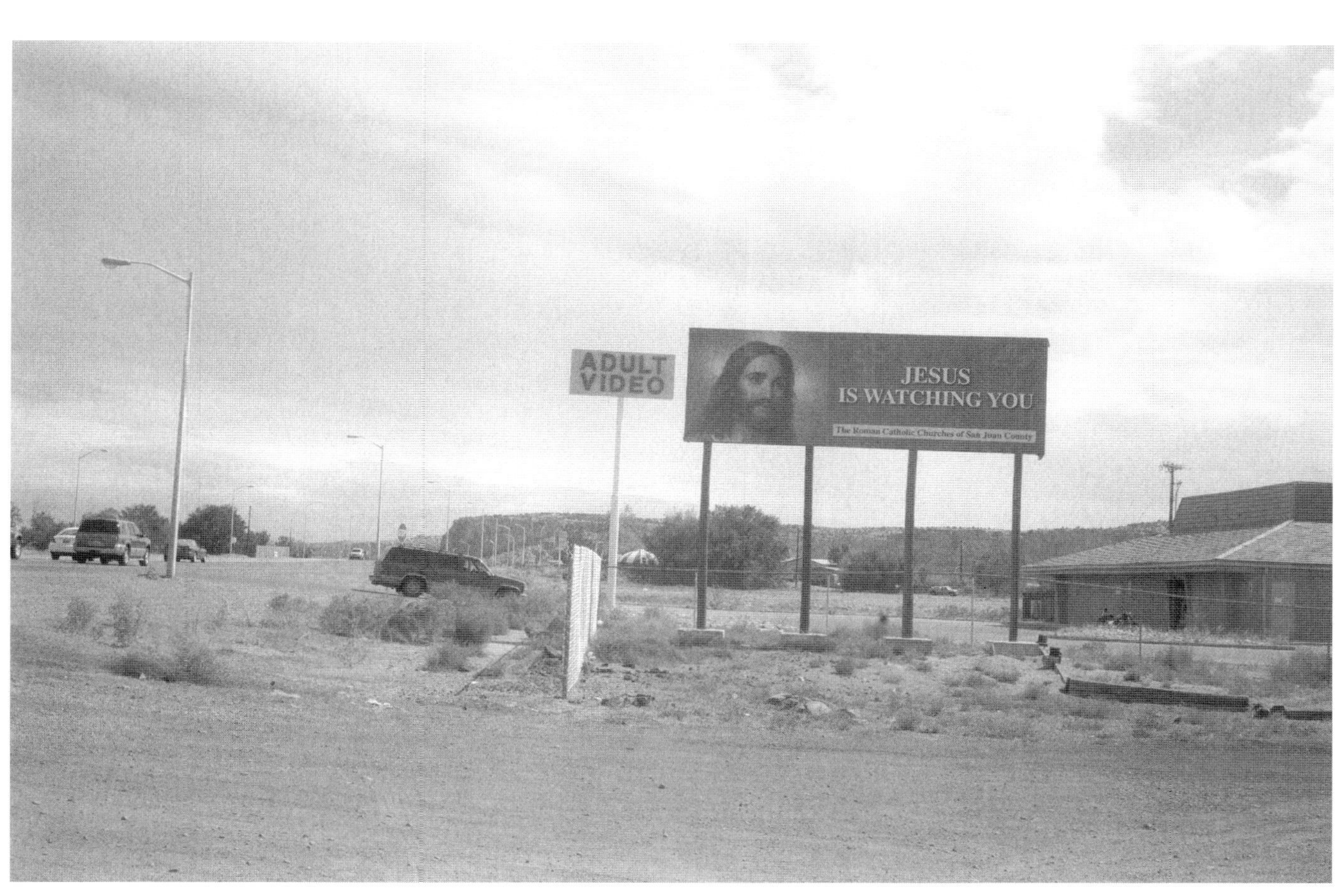

ADULT
VIDEO
JESUS
IS WATCHING YOU
The Roman Catholic Churches of San Juan County

19   ❖   Attention grouse hunters

Highway 14, Poudre River Valley, Colorado, 2000

While the science of it all is understandable, the public relations aspect needs some refining. Using the word *harvested* definitely sits better with non-hunters—makes a person think of a farmer bringing in the crops on a cool morning—but a billboard, fifty-five gallon drum, and graphic cutting instructions pretty much overrides that benefit. I wonder what they do during deer season?

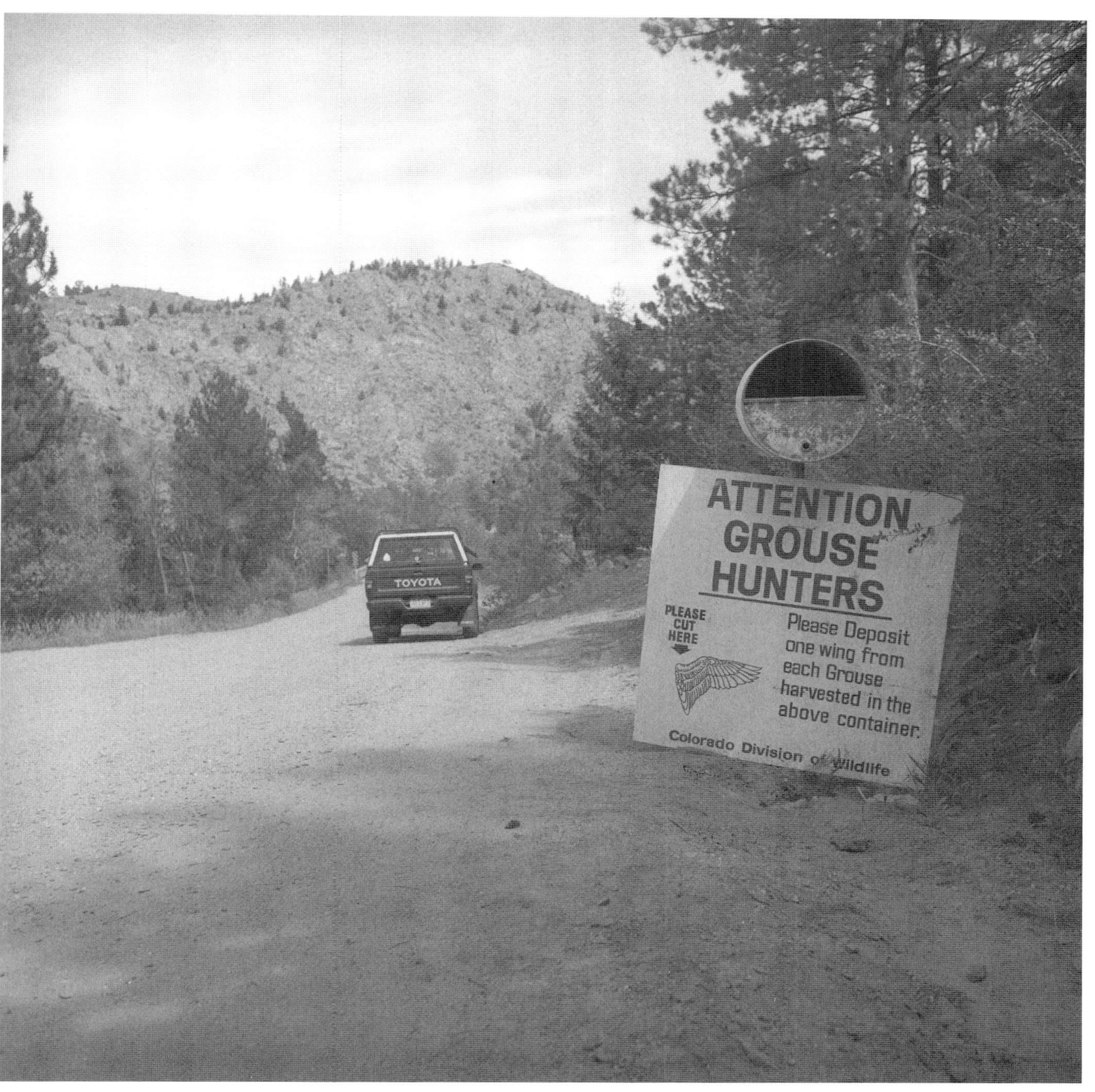
TOYOTA
ATTENTION
GROUSE
HUNTERS
PLEASE
CUT
HERE
Please Deposit
one wing from
each Grouse
harvested in the
above container.
Colorado Division of Wildlife

Highway 56, east of Dodge City, Kansas, 2000

Overheard conversation in the observation gazebo:
    Woman: "Are those the wagon tracks?"
    Man #1: "I don't think so, looks more like a four-wheeler, probably maintenance."
    Man #2: "Naw, those are pickup tracks, I'd guess a Ford."

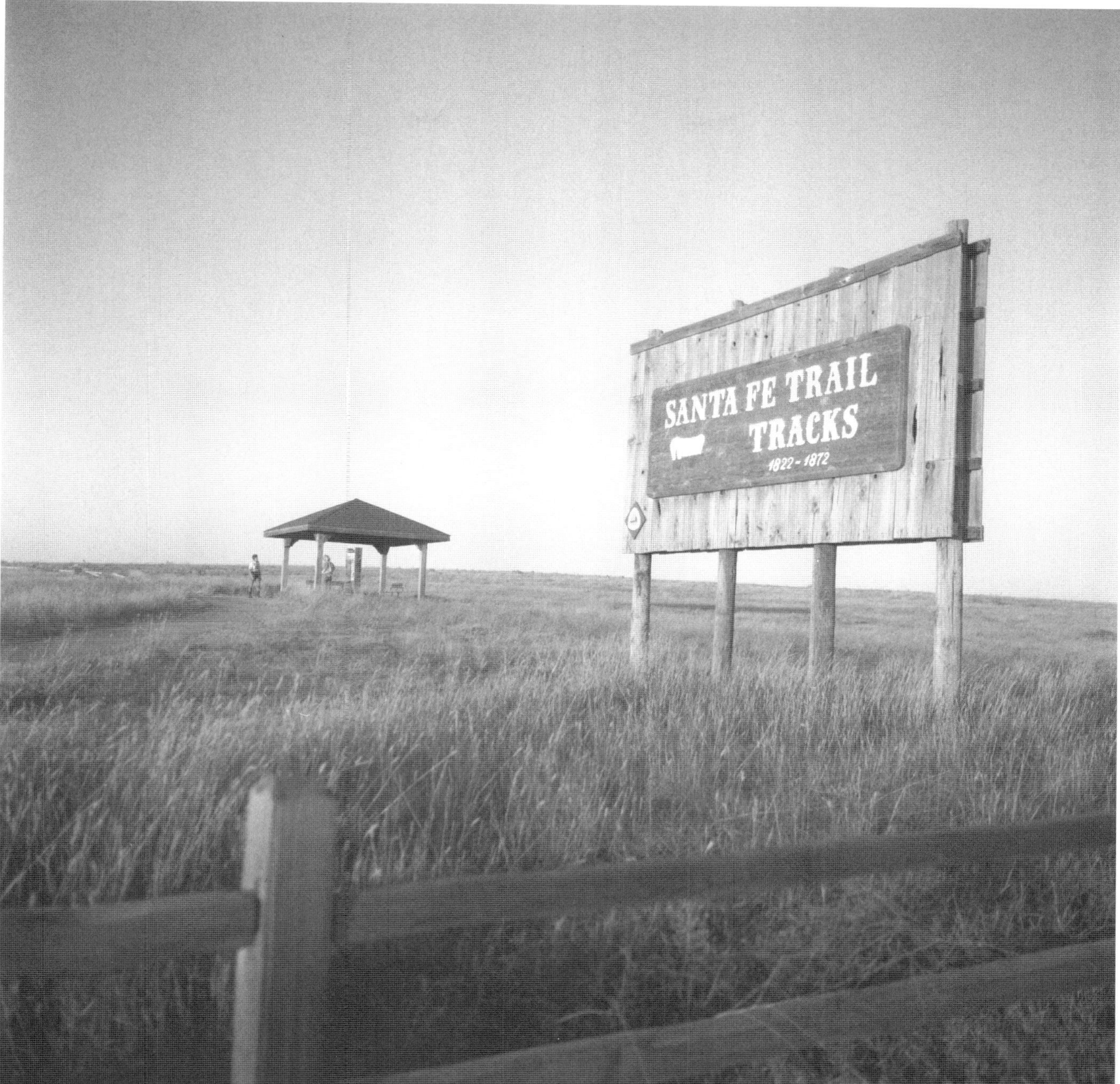

SANTA FE TRAIL
TRACKS
1822 - 1872

❖   Hugs–N–Tugs

Highway 412, near Hardy, Arkansas, 2008

Traffic was heavy when I tried to stop for this one. I had to circle around four or five times before I could get off the highway and into position. I noticed several people staring out of the daycare windows at me. Normally I would have walked over, knocked on the door, and explained what I was up to, but I was looking rough from a week of travel and figured they probably wouldn't open the door for a man circling around with a camera, anyway. So, I just got back in the truck and drove off, expecting to see blue lights at any moment, wondering if the humor of odd sign placement would be apparent to a Sharp County sheriff's deputy.

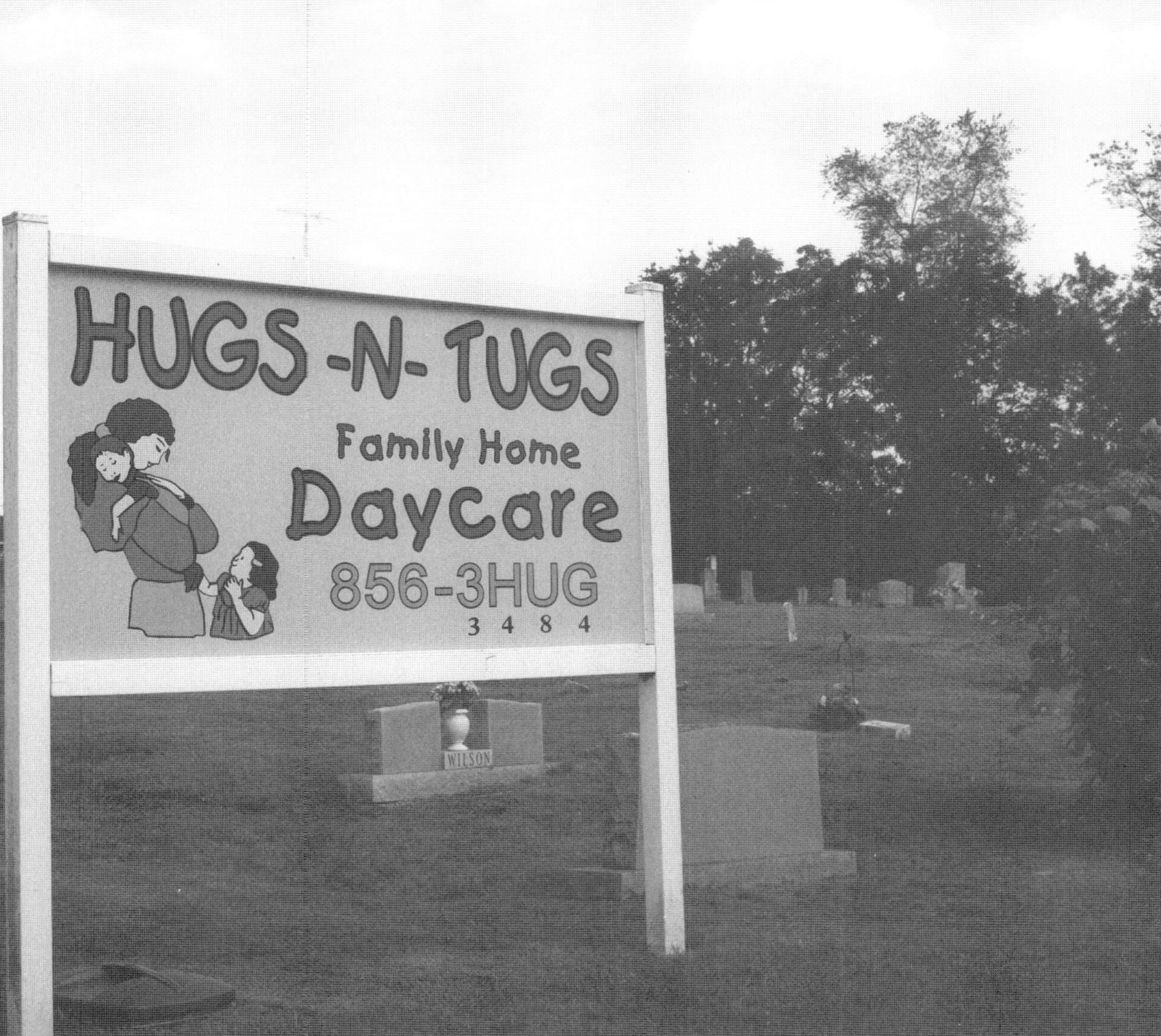
HUGS -N- TUGS
Family Home
Daycare
856-3HUG
3 4 8 4
WILSON

❖ Flippin Church of God

Highway 412, Flippin, Arkansas, 2005

Who could have resisted this?

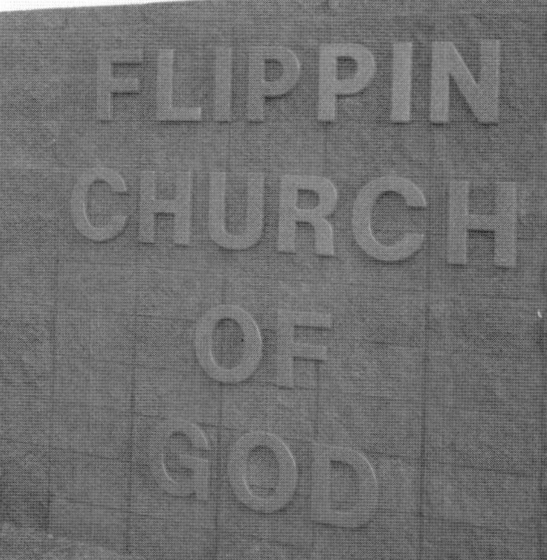
FLIPPIN
CHURCH
OF
GOD

❖ Keystone Gallery

Highway 83, Monument Rocks National Landmark, Kansas, 2000

Okay, we only have enough money to repaint one faded sign. Should we do the one that warns people of the extreme danger of falling rocks or the one that points them toward the gift shop?

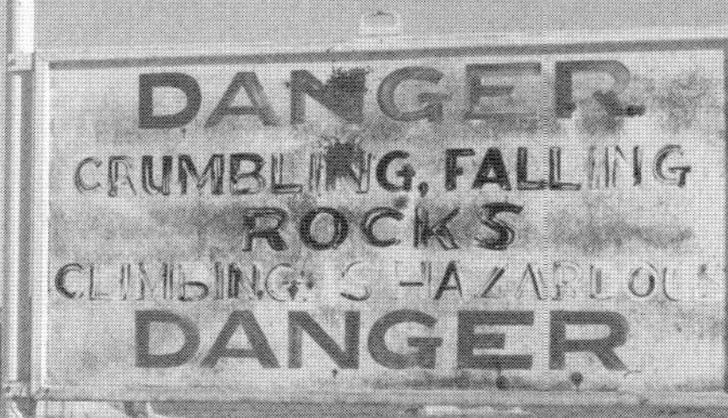
DANGER
CRUMBLING, FALLING
ROCKS
CLIMBING IS HAZARDOUS
DANGER
VISIT
KEYSTONE GALLERY
FOSSIL MUSEUM
ARTWORK - GIFTS
2½ MILES S. - 6 MILES W. TO U.S. 83 HW.

24 ❖ Only a fool

Highway 24, Little Wild Horse Canyon, Utah, 2007

This is not your average federal agency trailhead sign. This is clearly the language of someone who is fed up with the stupidity of the general public.

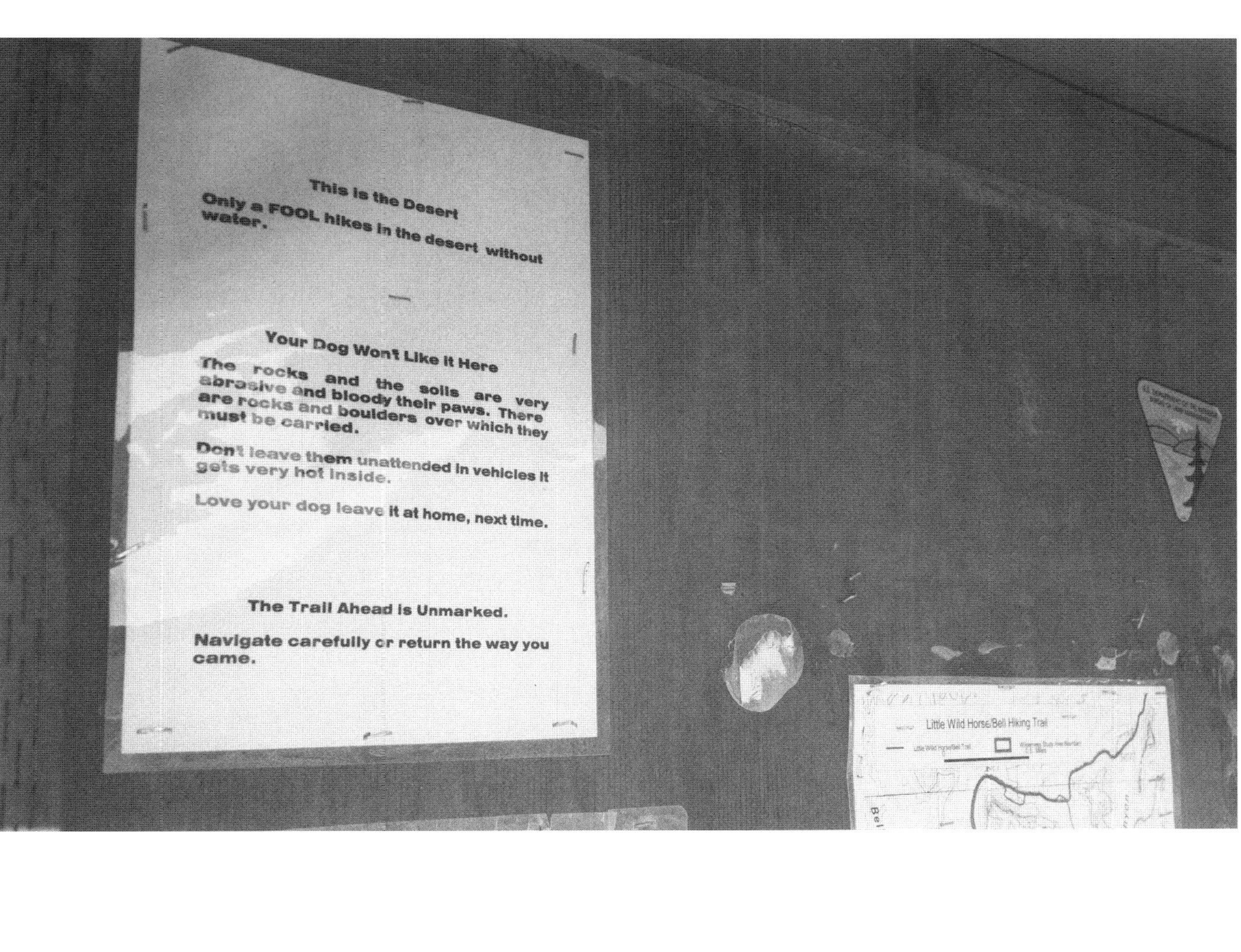
This is the Desert
Only a FOOL hikes in the desert without water.

Your Dog Won't Like It Here
The rocks and the soils are very abrasive and bloody their paws. There are rocks and boulders over which they must be carried.

Don't leave them unattended in vehicles it gets very hot inside.

Love your dog leave it at home, next time.

The Trail Ahead is Unmarked.

Navigate carefully or return the way you came.

Little Wild Horse/Bell Hiking Trail

❖   American owned

Highway 5, Ava, Missouri, 2008

Attach a sign that says *American Made,* and I'll go out of my way to buy it. Attach a sign that says *American Owned,* and I'll go out of my way to avoid it. One speaks of pride and one speaks of prejudice.

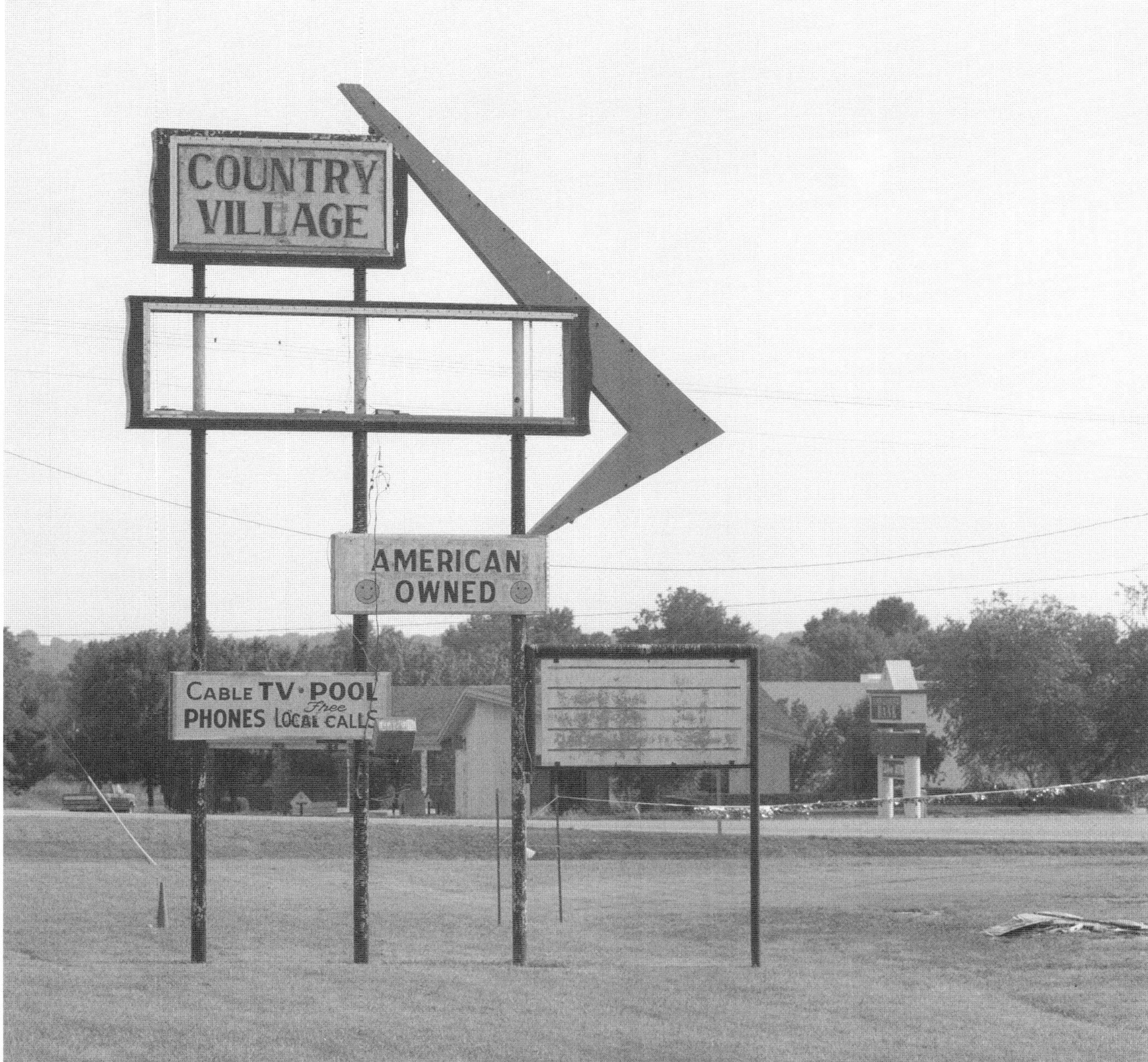

COUNTRY VILLAGE
AMERICAN OWNED
CABLE TV · POOL
PHONES Free LOCAL CALLS

Designated construction site

Highway 50, Clermont, Florida, 2003

The flag of Florida should show a beautiful palm tree with a bulldozer pushing it over. The level of development that pervades much of the state can evoke thoughts of disease; words like epidemic, cancerous, fungus come to mind. This spot, near Orlando, is intriguing because of the word *designated,* something usually associated with wetlands and wildlife sanctuaries. Leave it to Florida to prioritize the protection of a construction site.

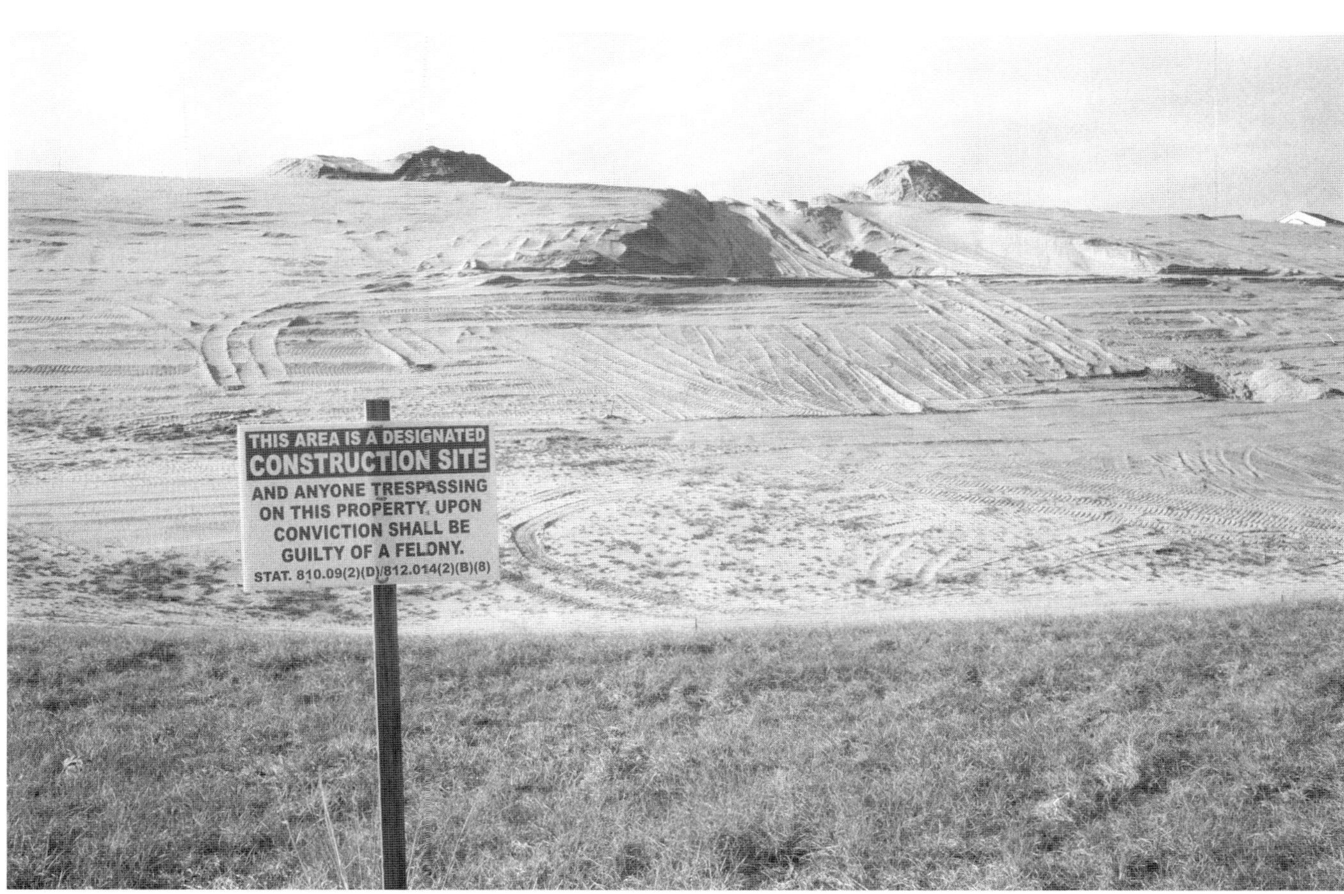

THIS AREA IS A DESIGNATED
CONSTRUCTION SITE
AND ANYONE TRESPASSING
ON THIS PROPERTY, UPON
CONVICTION SHALL BE
GUILTY OF A FELONY.
STAT. 810.09(2)(D)/812.014(2)(B)(8)

27

❖   Corn hole games

Highway 725, near Camden, Ohio, 2008

I know it's immature of me, but I found myself giggling through most of western Ohio. When I was growing up, corn hole was a vulgar slang for rectum. Corn hole games? The mind boggles.

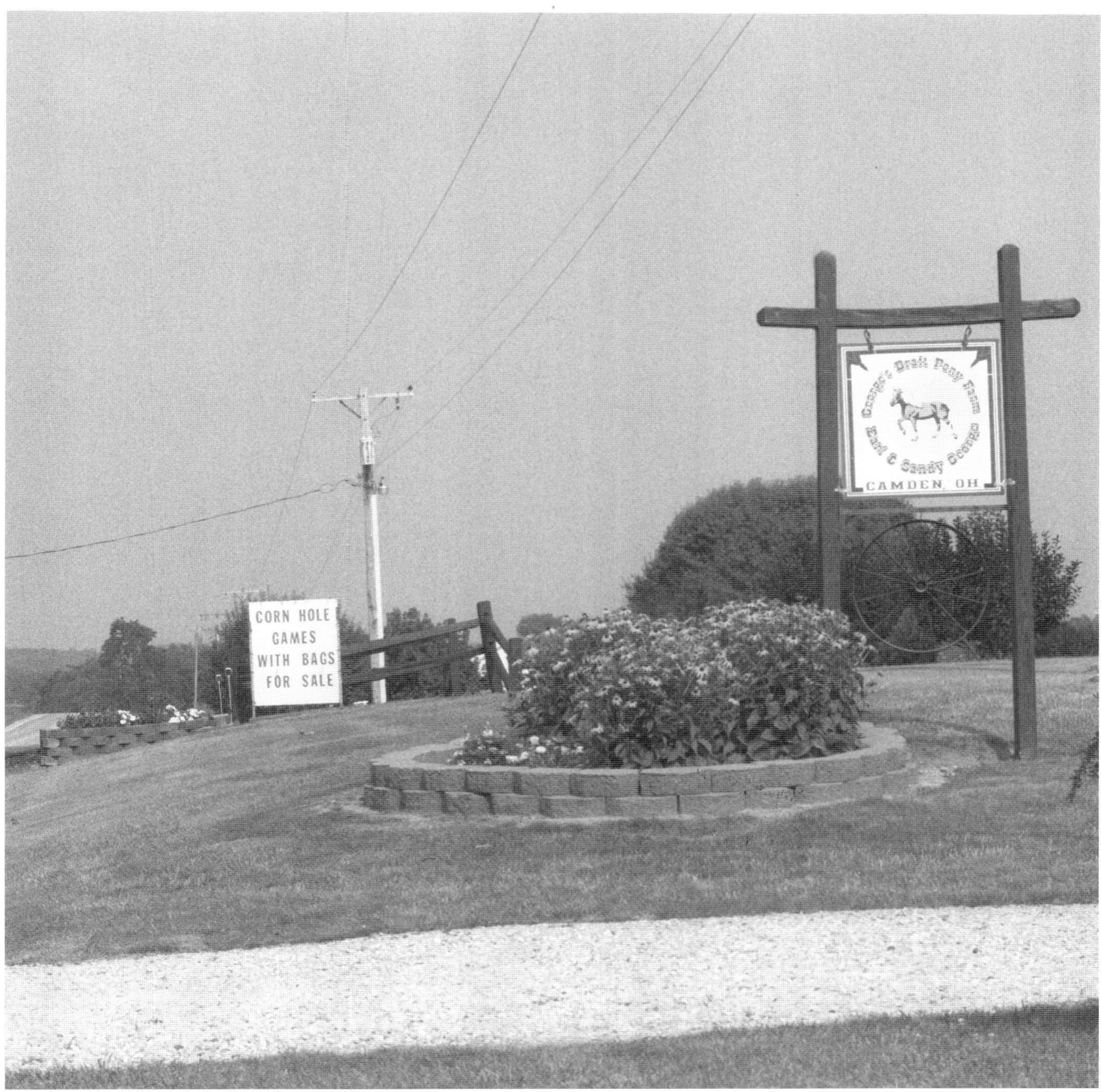
George's Draft Pony Farm
Earl & Sandy George
CAMDEN, OH
CORN HOLE
GAMES
WITH BAGS
FOR SALE

❖ Chritmas break

Highway 16, Elkins, Arkansas, 2002

The inspiration for the title of this book.

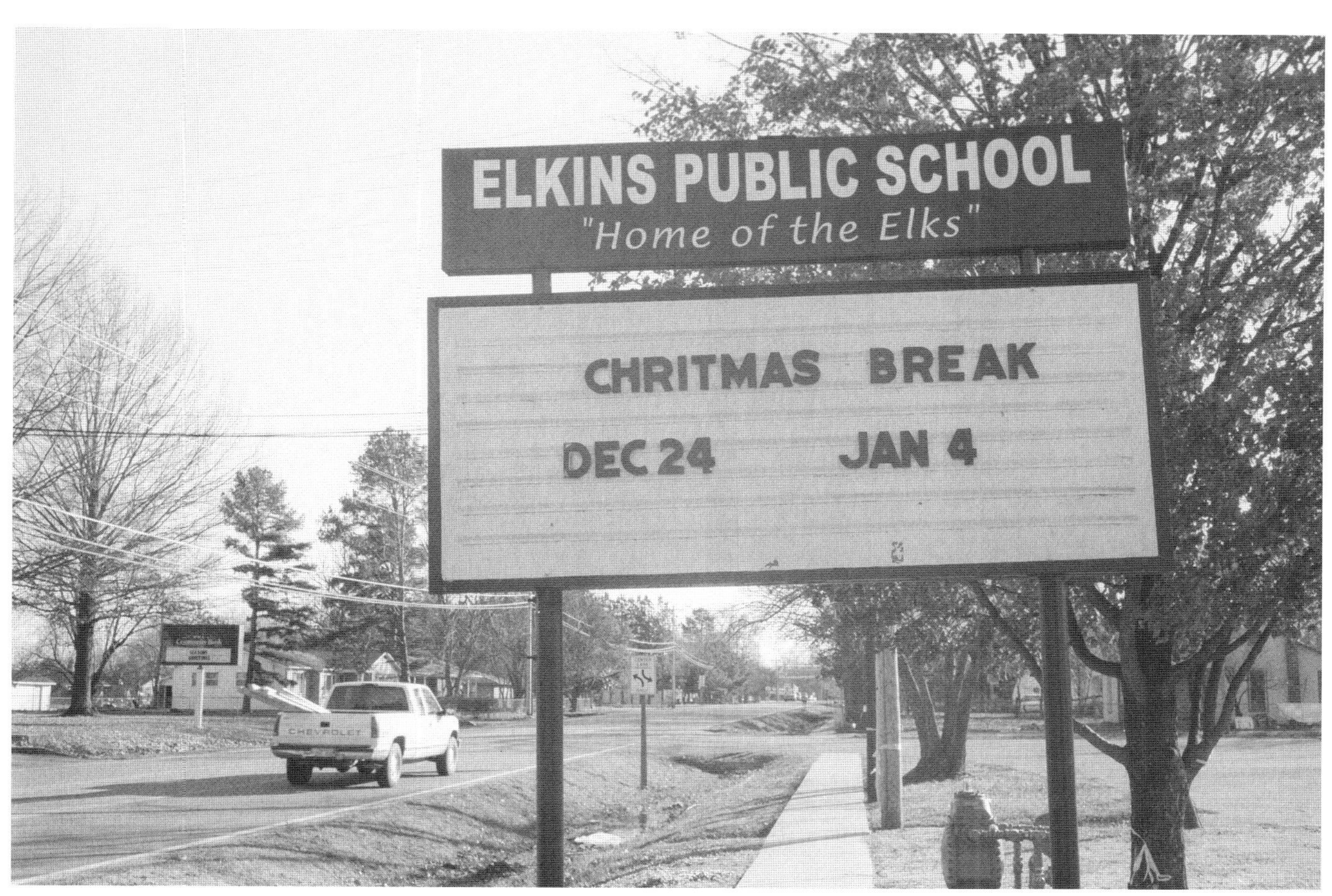

ELKINS PUBLIC SCHOOL
"Home of the Elks"
CHRITMAS BREAK
DEC 24   JAN 4
CHEVROLET

❖   Hillary Clinton

Highway 54, near Mullinville, Kansas, 2000

Images of Hillary Clinton and Janet Reno and some rough descriptions of their characters were what made me pull to the side of the road. I was impressed with the sheer magnitude of the display, the amount of work it represented, but I couldn't help imagining discussions that must have taken place at the Mullinville Chamber of Commerce.

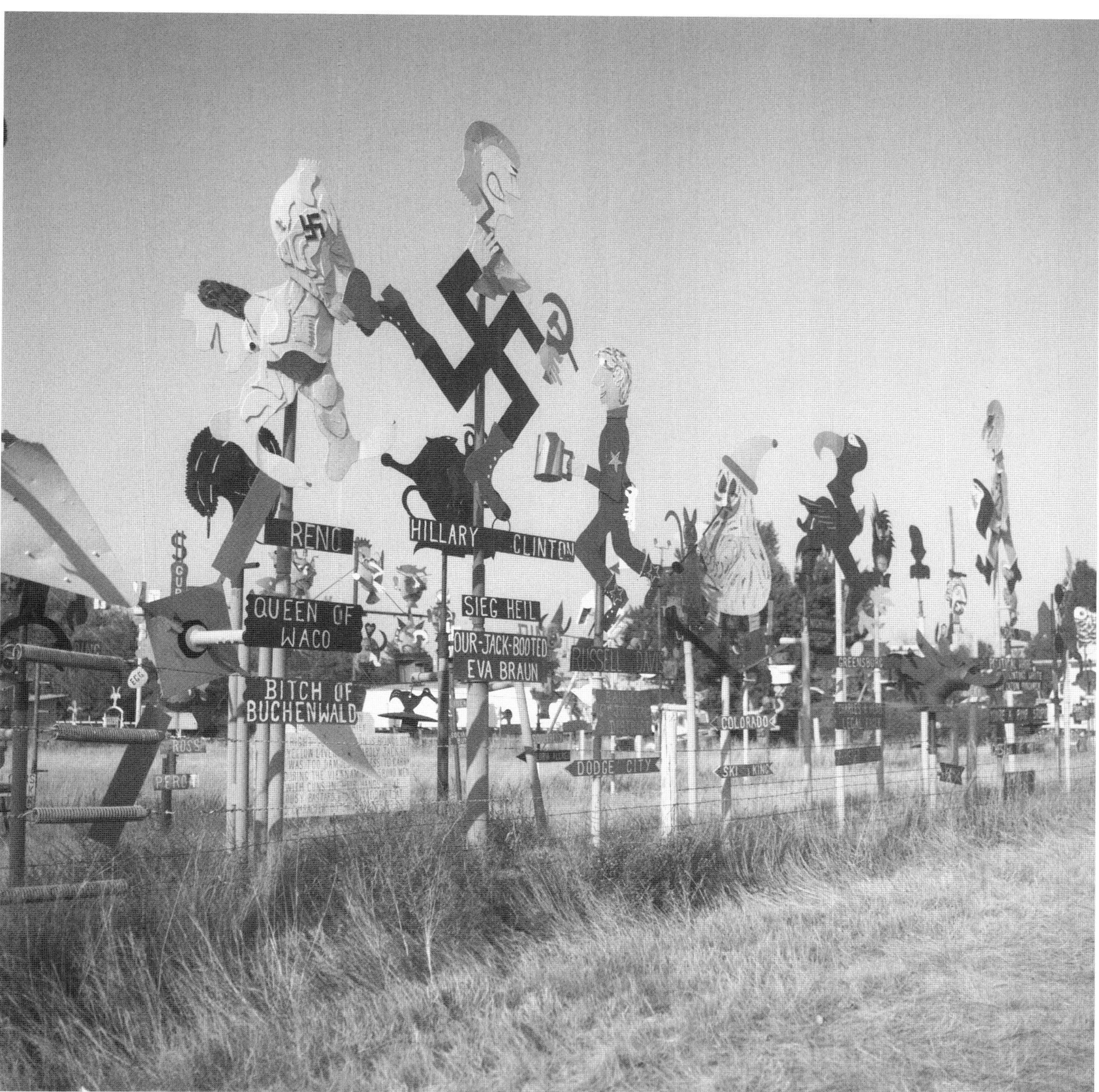
RENO
HILLARY CLINTON
QUEEN OF WACO
SIEG HEIL
OUR-JACK-BOOTED
EVA BRAUN
BITCH OF BUCHENWALD
RUSSELL DAVIS
GREENSBURG
COLORADO
DODGE CITY
SKI KING
PEROT

30

❖ Holy evolution

Highway 54, near Mullinville, Kansas, 2007

Returning seven years later, the display had expanded down the highway. People were stopping their cars, getting out and stumbling along the ditch past signs of humor, anger, cynicism, sarcasm, disgust, and disdain with all things political. Art as we know it may have started just like this.

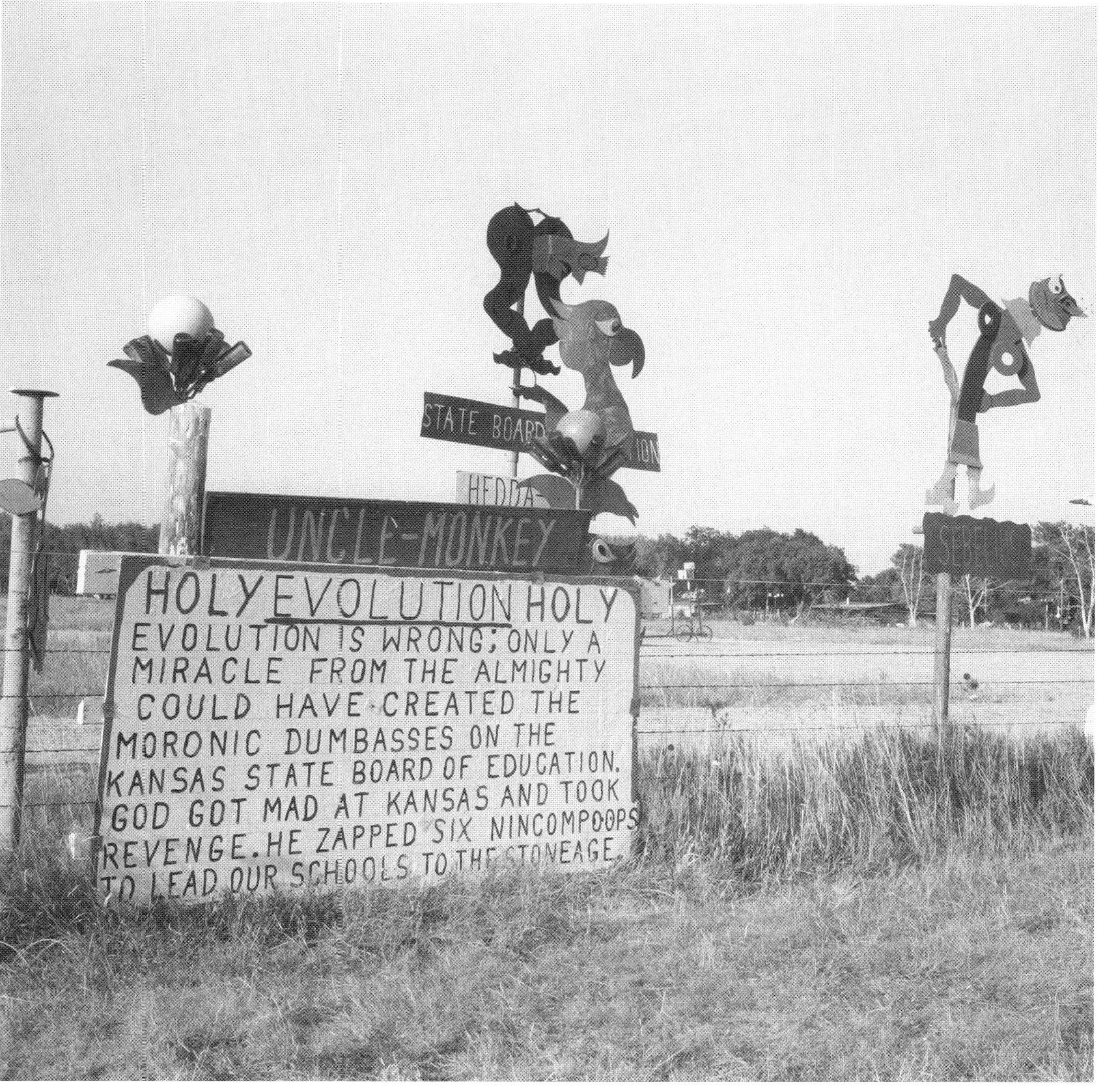
STATE BOARD
ION
HEDDA-
UNCLE-MONKEY
HOLY EVOLUTION HOLY
EVOLUTION IS WRONG; ONLY A
MIRACLE FROM THE ALMIGHTY
COULD HAVE CREATED THE
MORONIC DUMBASSES ON THE
KANSAS STATE BOARD OF EDUCATION.
GOD GOT MAD AT KANSAS AND TOOK
REVENGE. HE ZAPPED SIX NINCOMPOOPS
TO LEAD OUR SCHOOLS TO THE STONE AGE.

❖   Farts store

Highway 16, Fayetteville, Arkansas, 2009

Just staring off into space waiting for the light to turn, and lightning strikes.

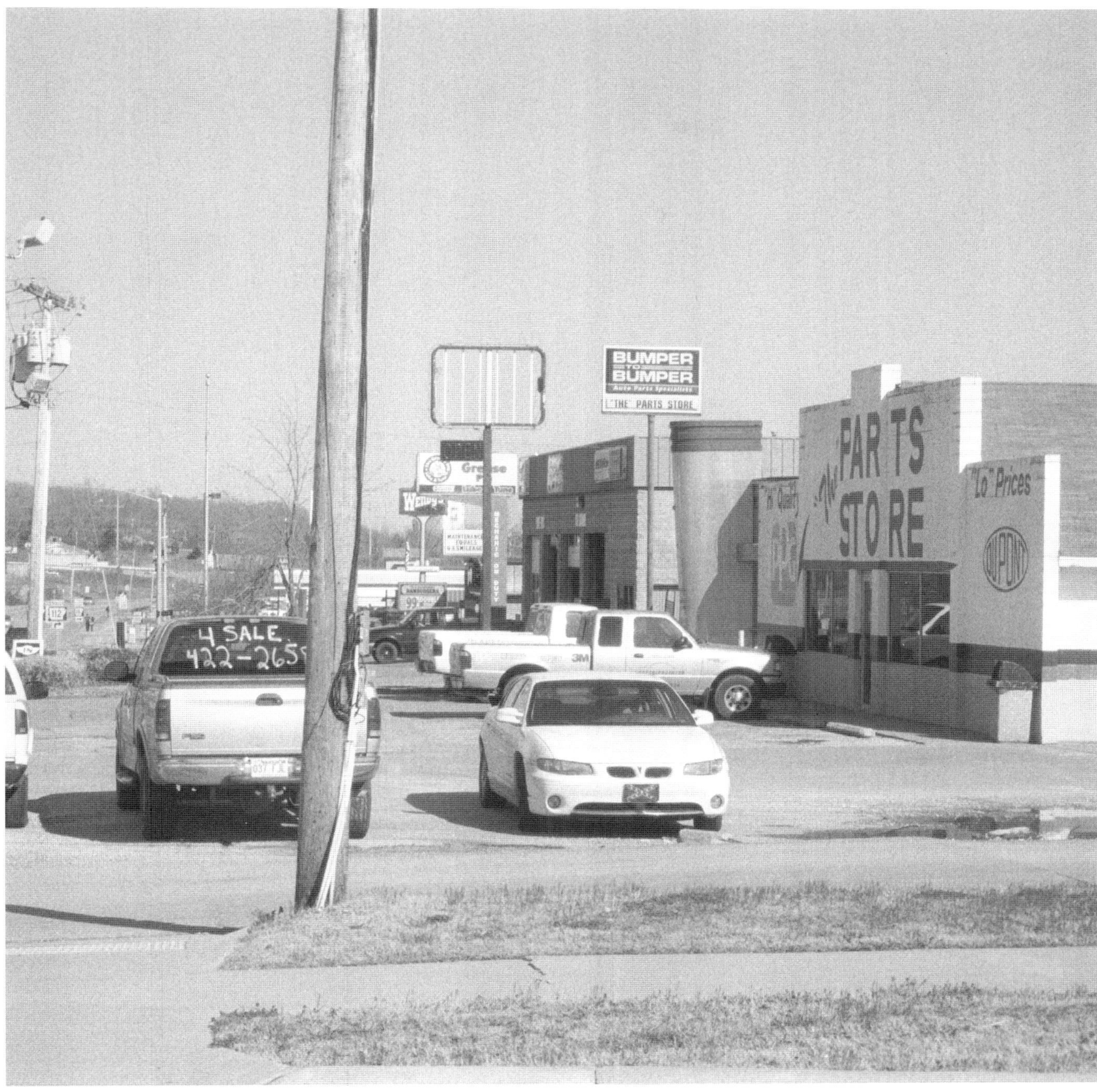

BUMPER TO BUMPER
Auto Parts Specialists
THE PARTS STORE
PARTS STORE
"Lo" Prices
DUPONT
Grease
Wendy's
4 SALE
422-265
3M
99¢

❖ Wildife conservation club

Highway 36, near Hamburg, Michigan, 2001

And afterward, there was a business meeting, highlighting the successes of the year, and pondering the one failure—an inexplicable decline in fish populations.

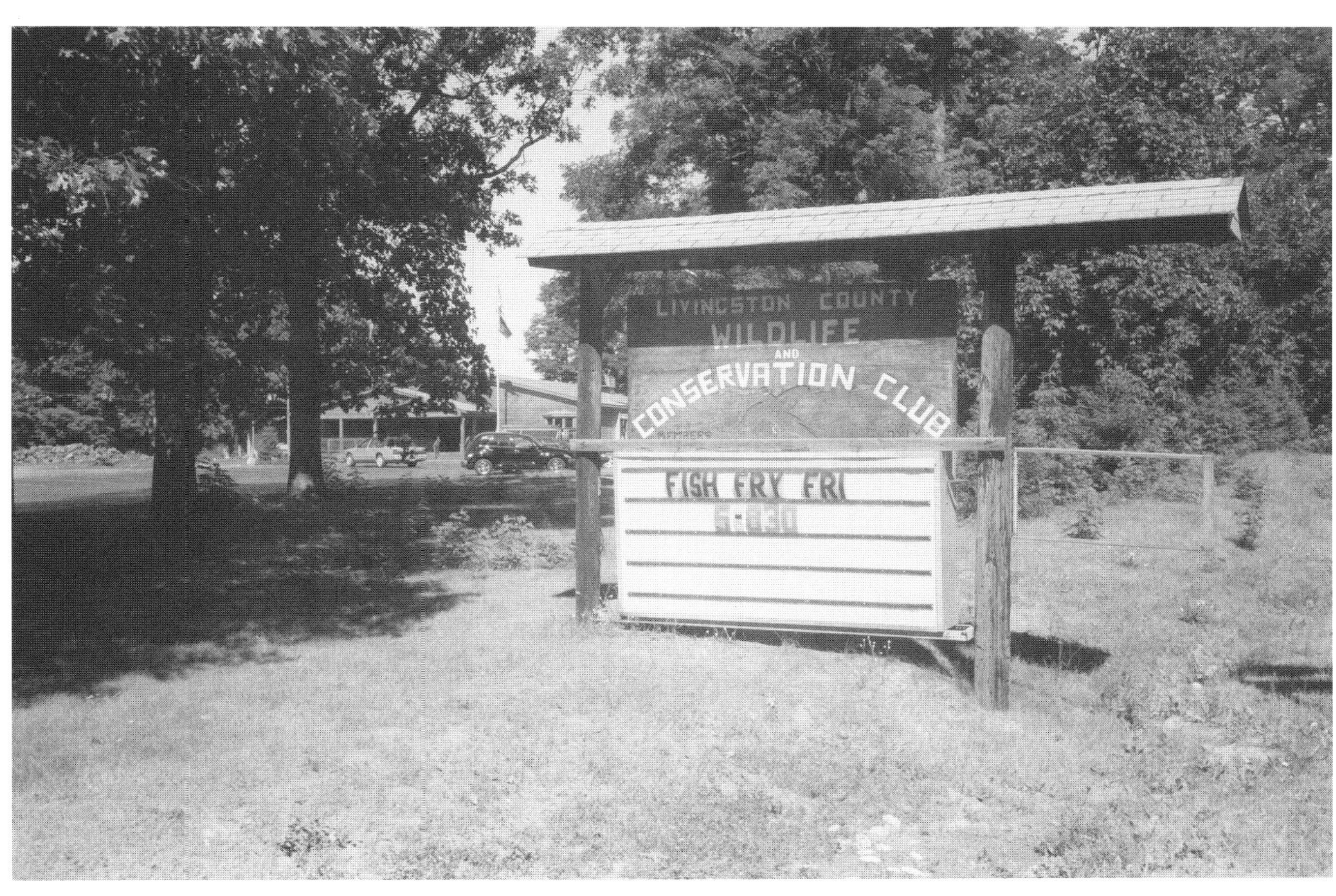

LIVINGSTON COUNTY
WILDLIFE
AND
CONSERVATION CLUB
FISH FRY FRI
5-8:30

Highway 24, near Capitol Reef National Park, Utah, 2007

Institutional sign budgets work like this: Use it or lose it at the end of the year.

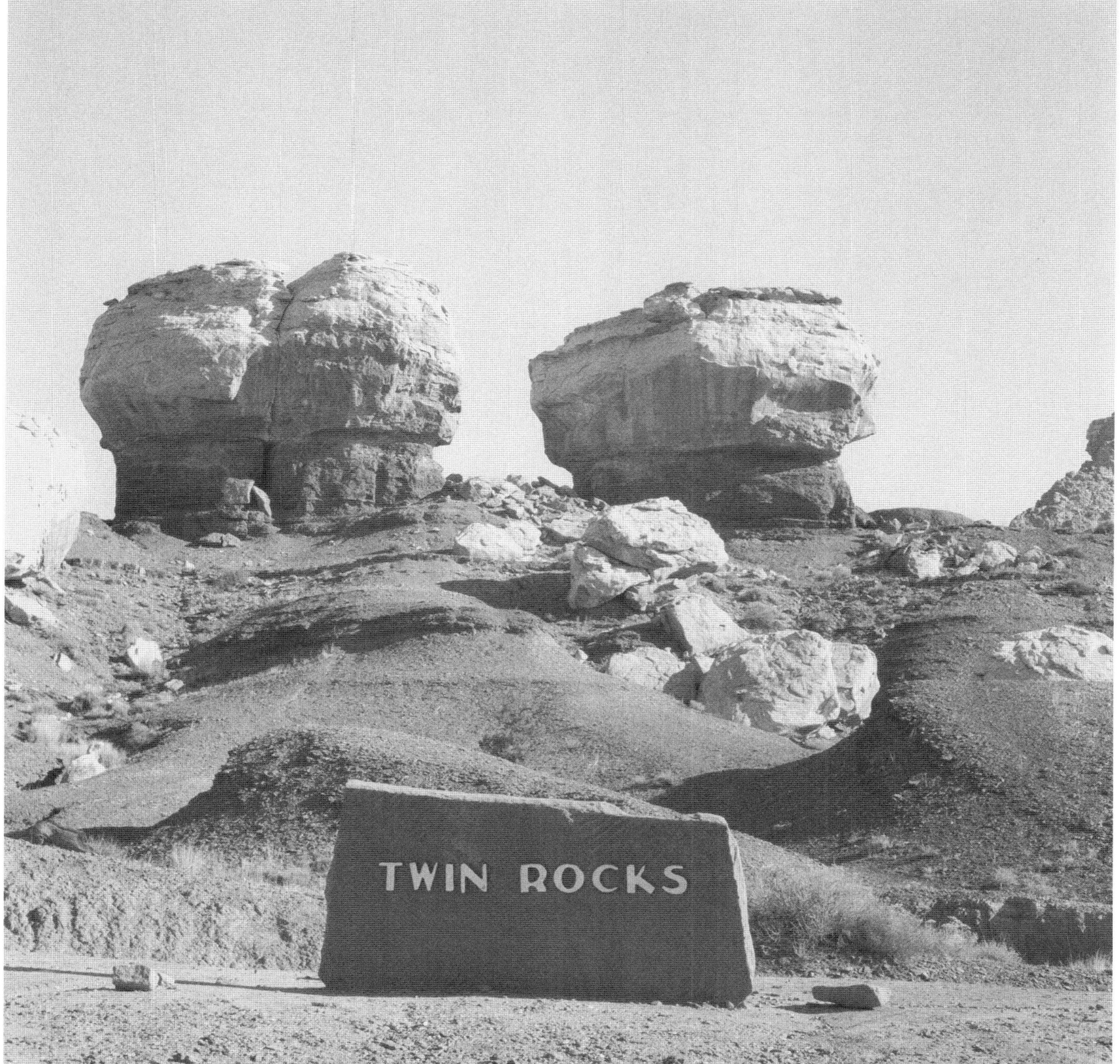
TWIN ROCKS

Highway 24, southern Utah, 2007

Just in case you thought that if you looked hard enough you could find this vista back home in Michigan.

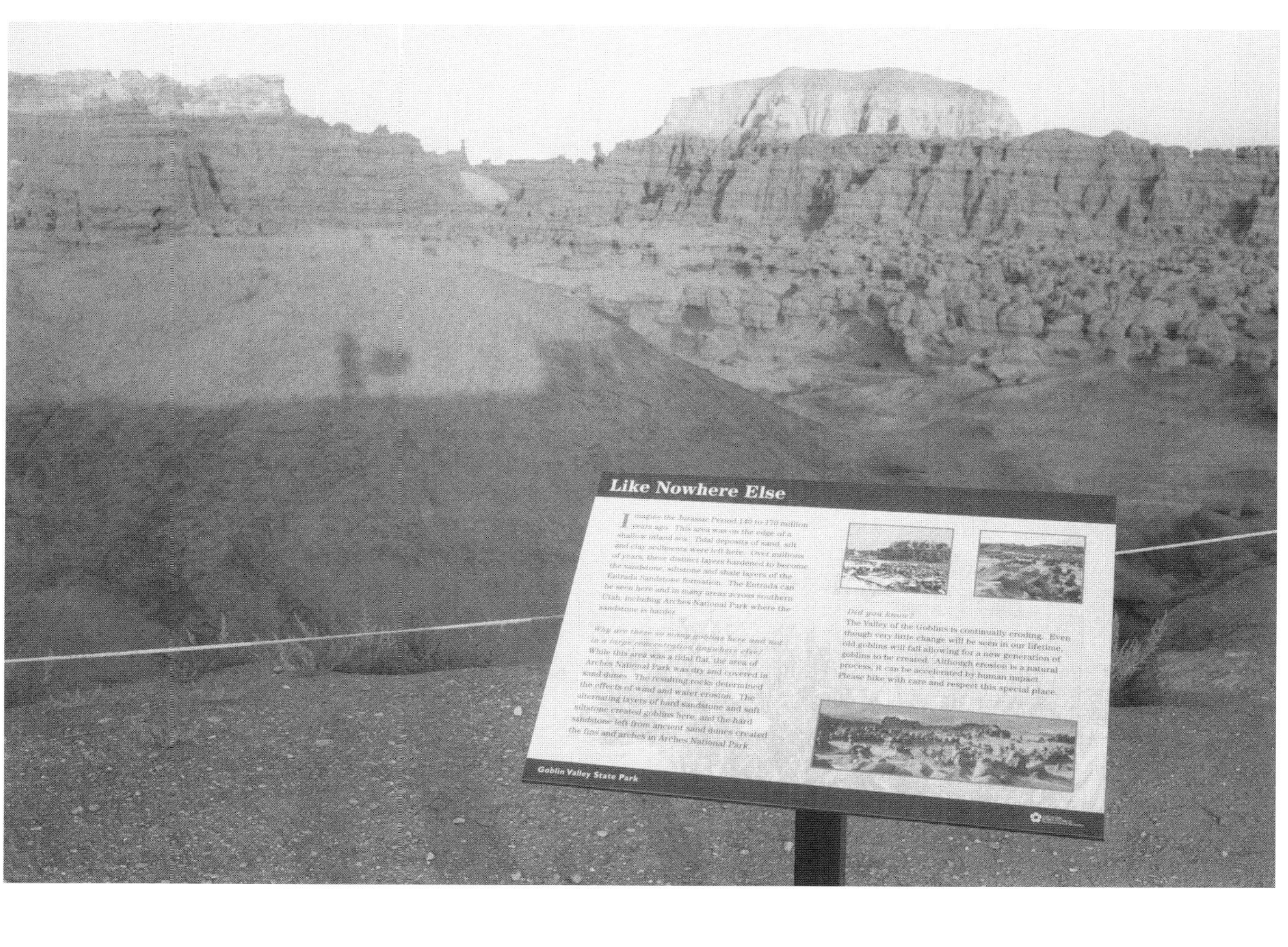
Like Nowhere Else
Imagine the Jurassic Period 140 to 170 million years ago. This area was on the edge of a shallow inland sea. Tidal deposits of sand, silt and clay sediments were left here. Over millions of years, these distinct layers hardened to become the sandstone, siltstone and shale layers of the Entrada Sandstone formation. The Entrada can be seen here and in many areas across southern Utah, including Arches National Park where the sandstone is harder.
Why are there so many goblins here and not in a large concentration anywhere else? While this area was a tidal flat, the area of Arches National Park was dry and covered in sand dunes. The resulting rocks determined the effects of wind and water erosion. The alternating layers of hard sandstone and soft siltstone created goblins here, and the hard sandstone left from ancient sand dunes created the fins and arches in Arches National Park.
Did you know?
The Valley of the Goblins is continually eroding. Even though very little change will be seen in our lifetime, old goblins will fall allowing for a new generation of goblins to be created. Although erosion is a natural process, it can be accelerated by human impact. Please hike with care and respect this special place.
Goblin Valley State Park

35

❖   Do not spit in ashtray

Highway 16, Elkins, Arkansas, 2002

This little corner of the family-owned BoMart grocery store in Elkins, Arkansas, just off Highway 16, has always fascinated me. The close proximity of the deadly (tobacco) and the benign (ice cream), the bulletin board crammed with business cards, lost animals, for sales, for rents, benefit suppers, and the ash can with its plea for civility. *God Bless America* turns it into an altar of sorts, a sacred place where all of us who pass by leave little representations of what's important.

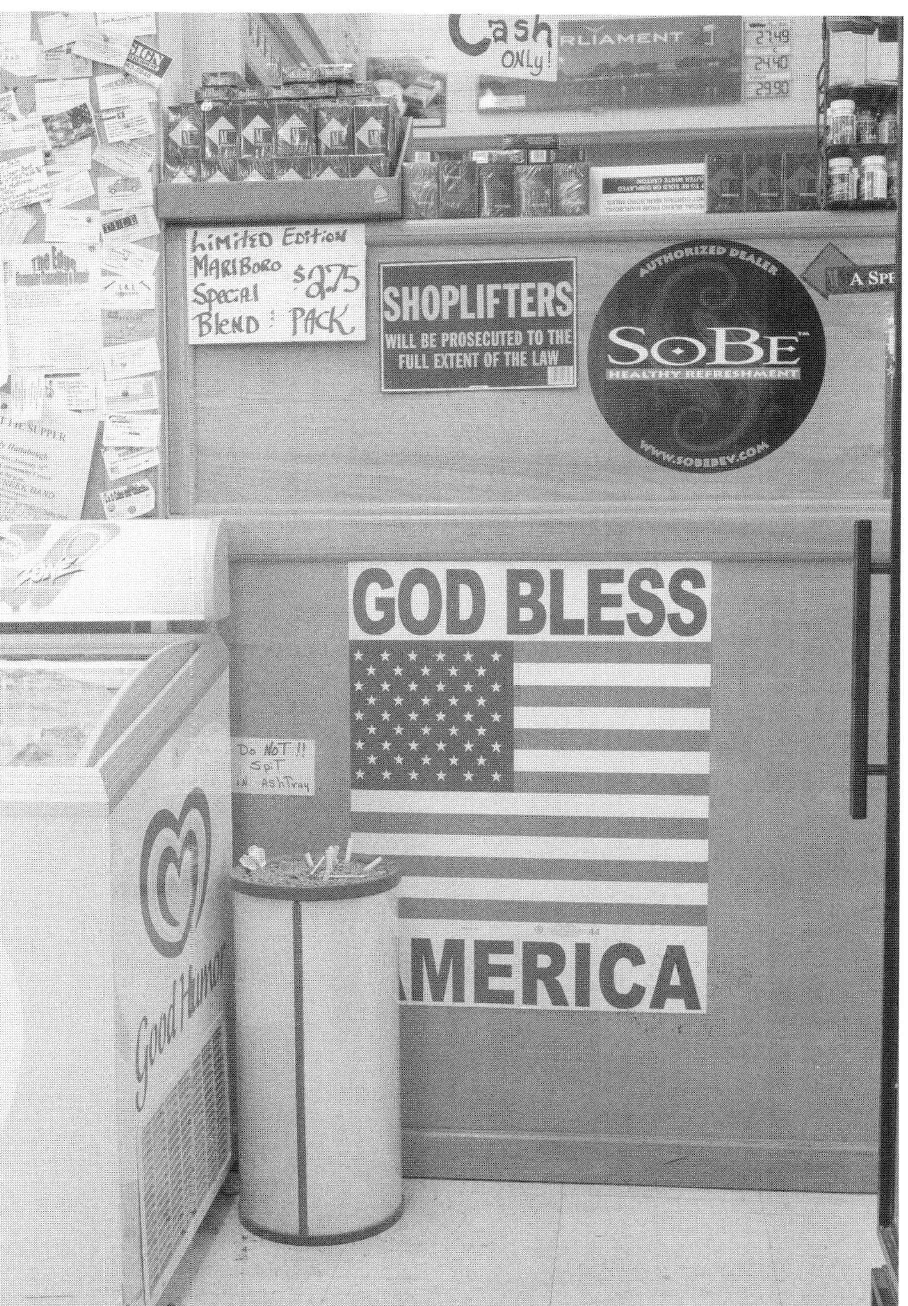
Cash ONLY!
RLIAMENT
27.49
24.40
29.90
Limited Edition
MARLBORO
Special
Blend : PACK
$2.75
SHOPLIFTERS
WILL BE PROSECUTED TO THE
FULL EXTENT OF THE LAW
AUTHORIZED DEALER
SoBe
HEALTHY REFRESHMENT
WWW.SOBEV.COM
A SPE
Do NOT !!
SpiT
iN AShTray
GOD BLESS
AMERICA
Good Humor

36

❖   Extreme danger

Highway 43, near Compton, Arkansas, Buffalo National River, 1992

As usual in a situation like this, it seems there are two possibilities:
   1) Go ahead, take a gamble.
   2) Turn back, and later regret it.
I chose #1, and experienced that bit of heaven known as the Buffalo National River, but later discovered that a young man had also recently chosen #1 and fallen to his death, before the sign was in place.

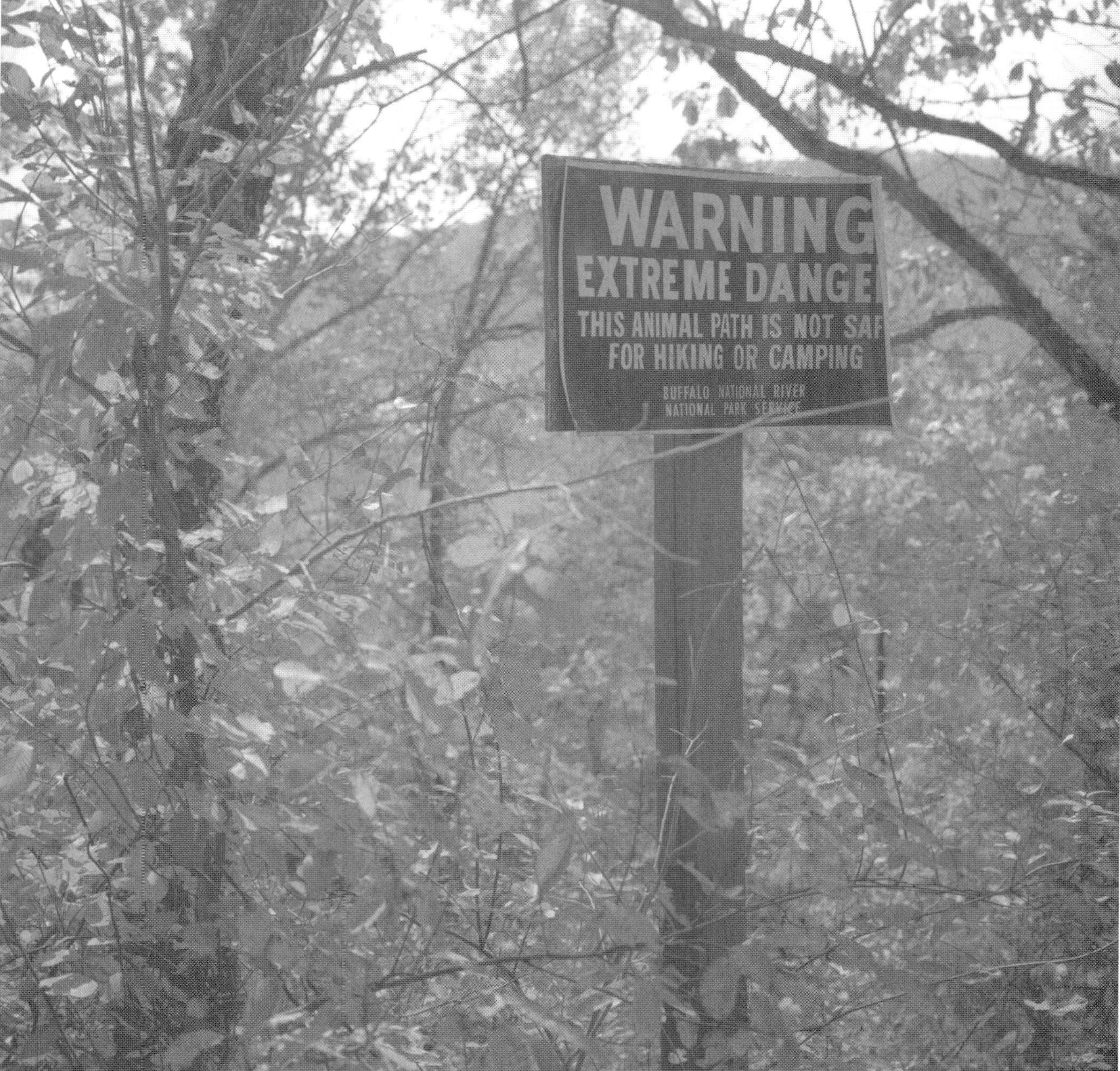

WARNING
EXTREME DANGER
THIS ANIMAL PATH IS NOT SAFE
FOR HIKING OR CAMPING
BUFFALO NATIONAL RIVER
NATIONAL PARK SERVICE

37

❖ Full gospel church

Highway 16, Elkins, Arkansas, 2001

I had driven by this sign for years before I realized it was a devious trap set by Satan to lure tired drivers to just follow the arrow to church.

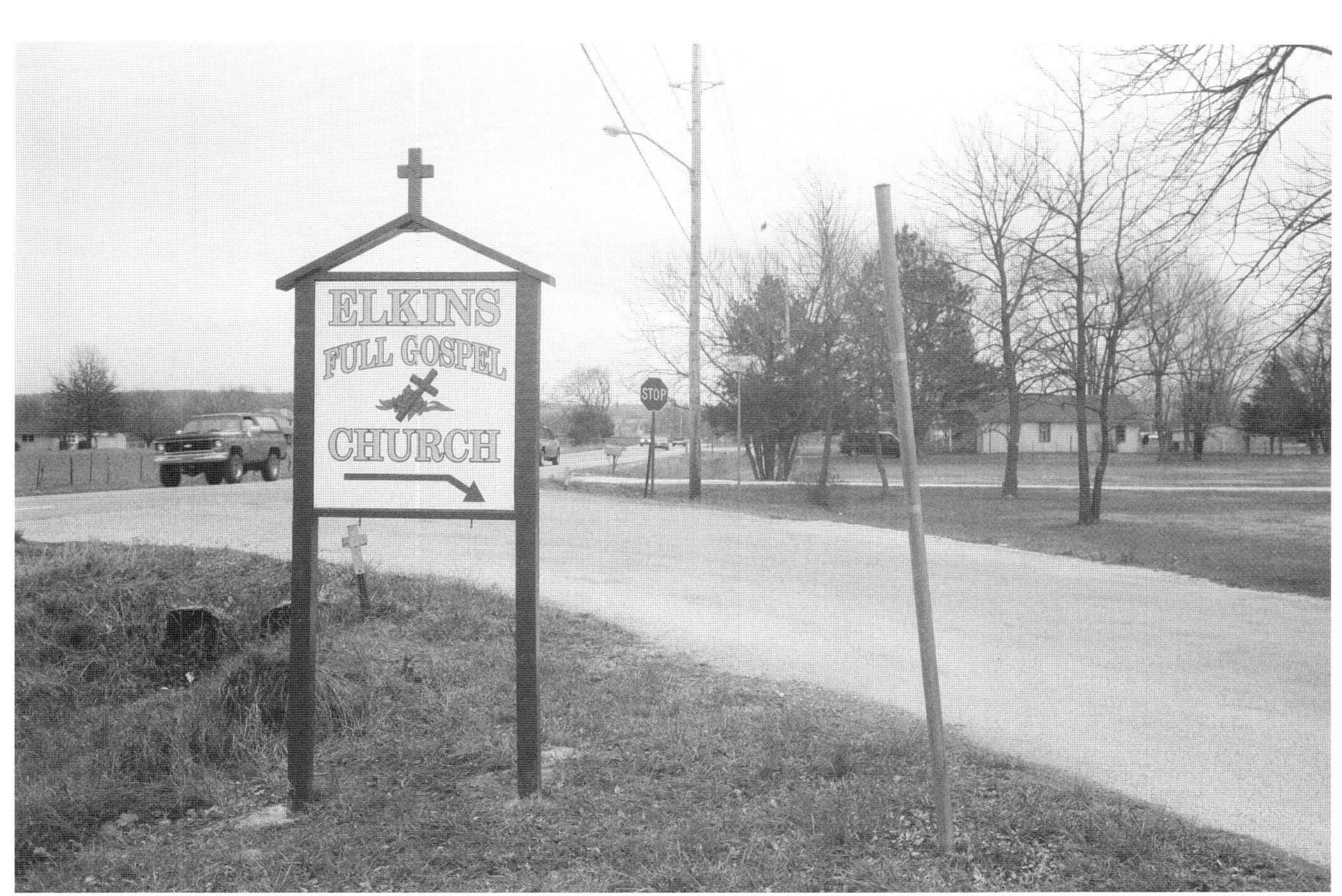

ELKINS
FULL GOSPEL
CHURCH
STOP

Highway 412, near Flippin, Arkansas, 2003

I have to admit I hadn't thought of *this* solution, but with the economic collapse of 2008 still in progress as this book goes to press, I'm reminded of Bob Seeger's lament... *wish I didn't know now what I didn't know then* (from the song "Running against the Wind").

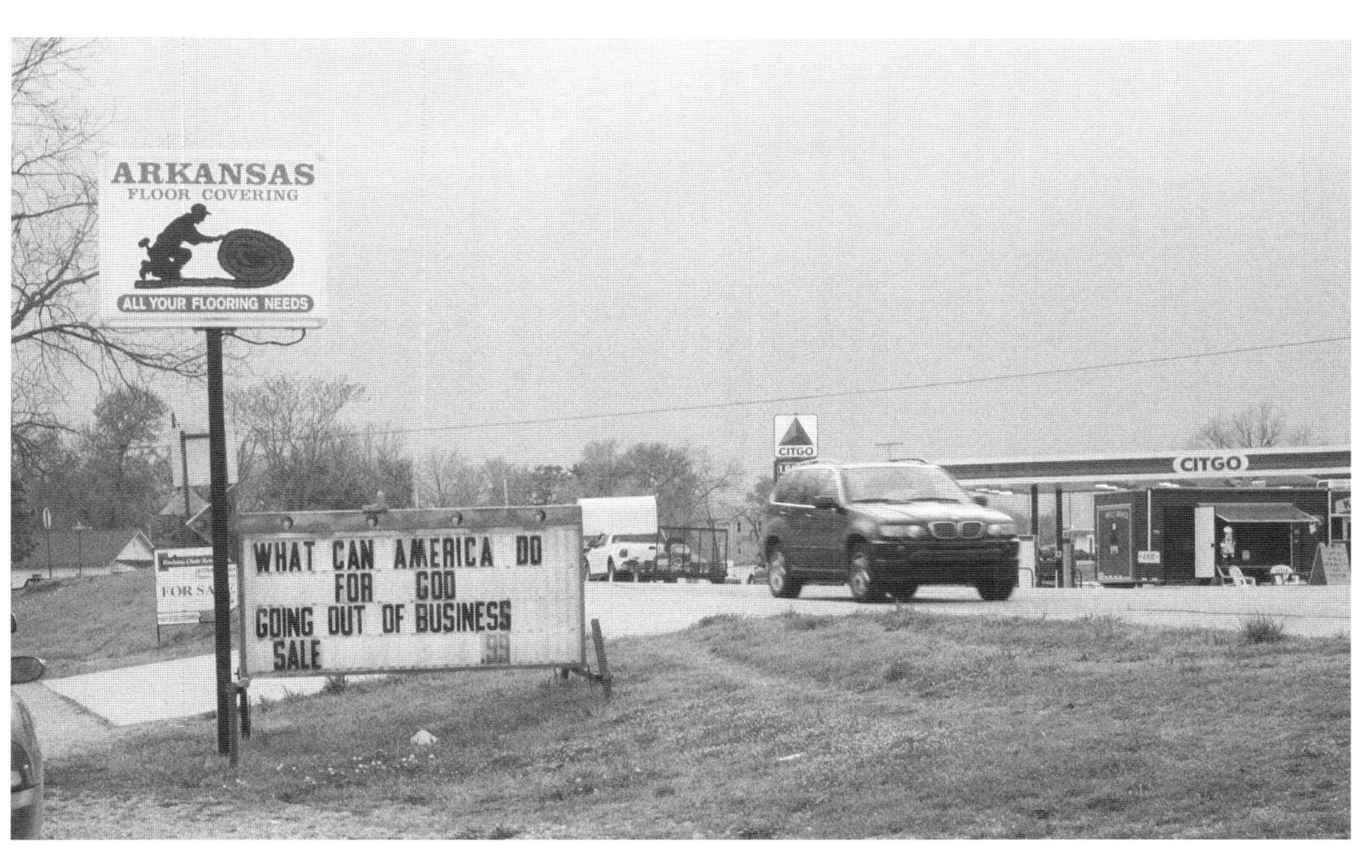

ARKANSAS
FLOOR COVERING
ALL YOUR FLOORING NEEDS
WHAT CAN AMERICA DO
FOR GOD
GOING OUT OF BUSINESS
SALE
FOR SA
CITGO
CITGO

39

❖   War is terrorism

Highway 45, near Goshen, Arkansas, 2003

Early 2003 found many Americans in a kind of depression. Someone in rural Arkansas snapped out of it long enough to try to do something.

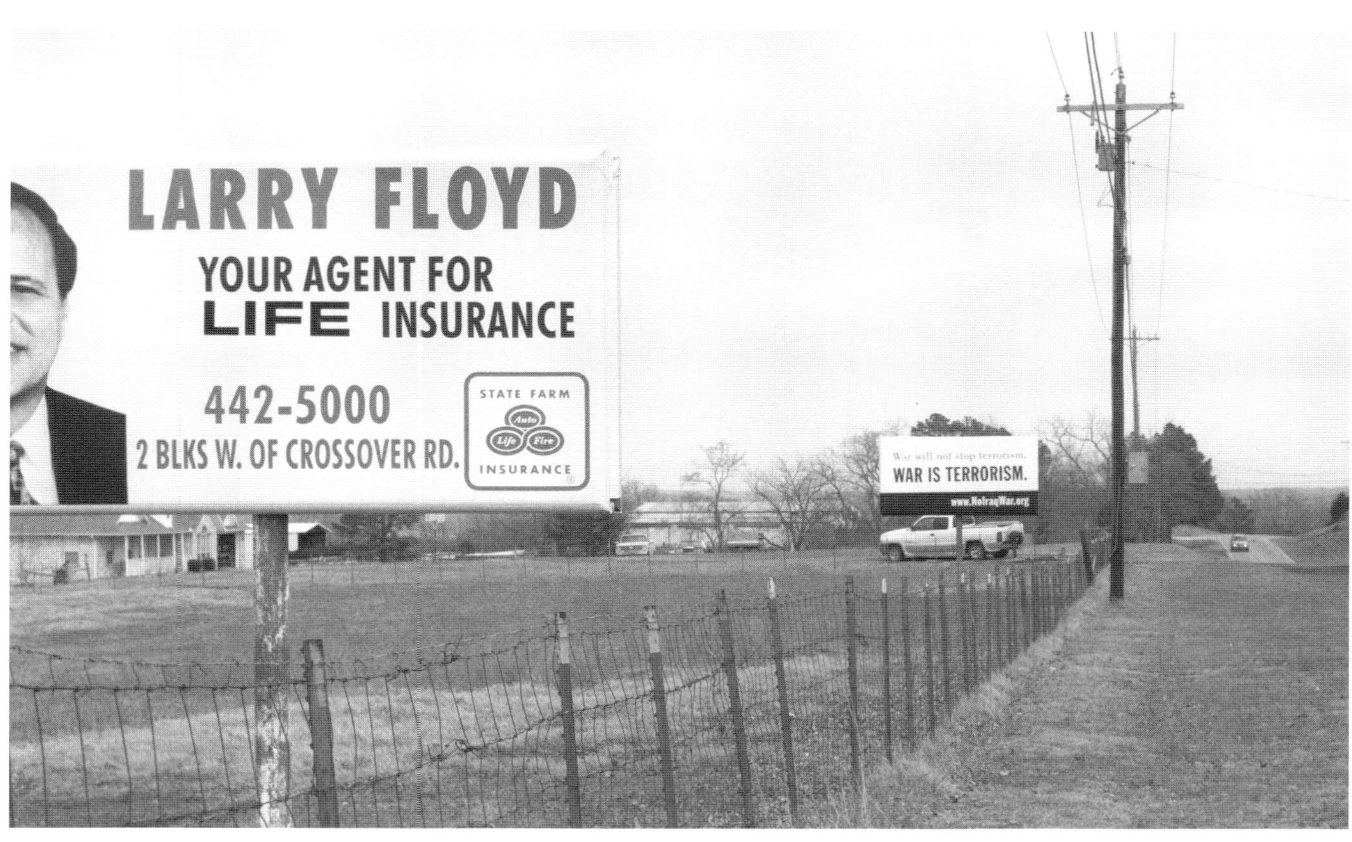

LARRY FLOYD
YOUR AGENT FOR
LIFE INSURANCE
442-5000
2 BLKS W. OF CROSSOVER RD.
STATE FARM
Auto Life Fire
INSURANCE
War will not stop terrorism.
WAR IS TERRORISM.
www.NoIraqWar.org

40

❖　　　No dumping allowed

Highway 71, near West Fork, Arkansas, 1998

Roadside memorials can be powerful reminders of the nearness of tragedy, the statistical game of chance we all play when we drive. The placement of this one seemed purposeful—a last act of defiance against authority.

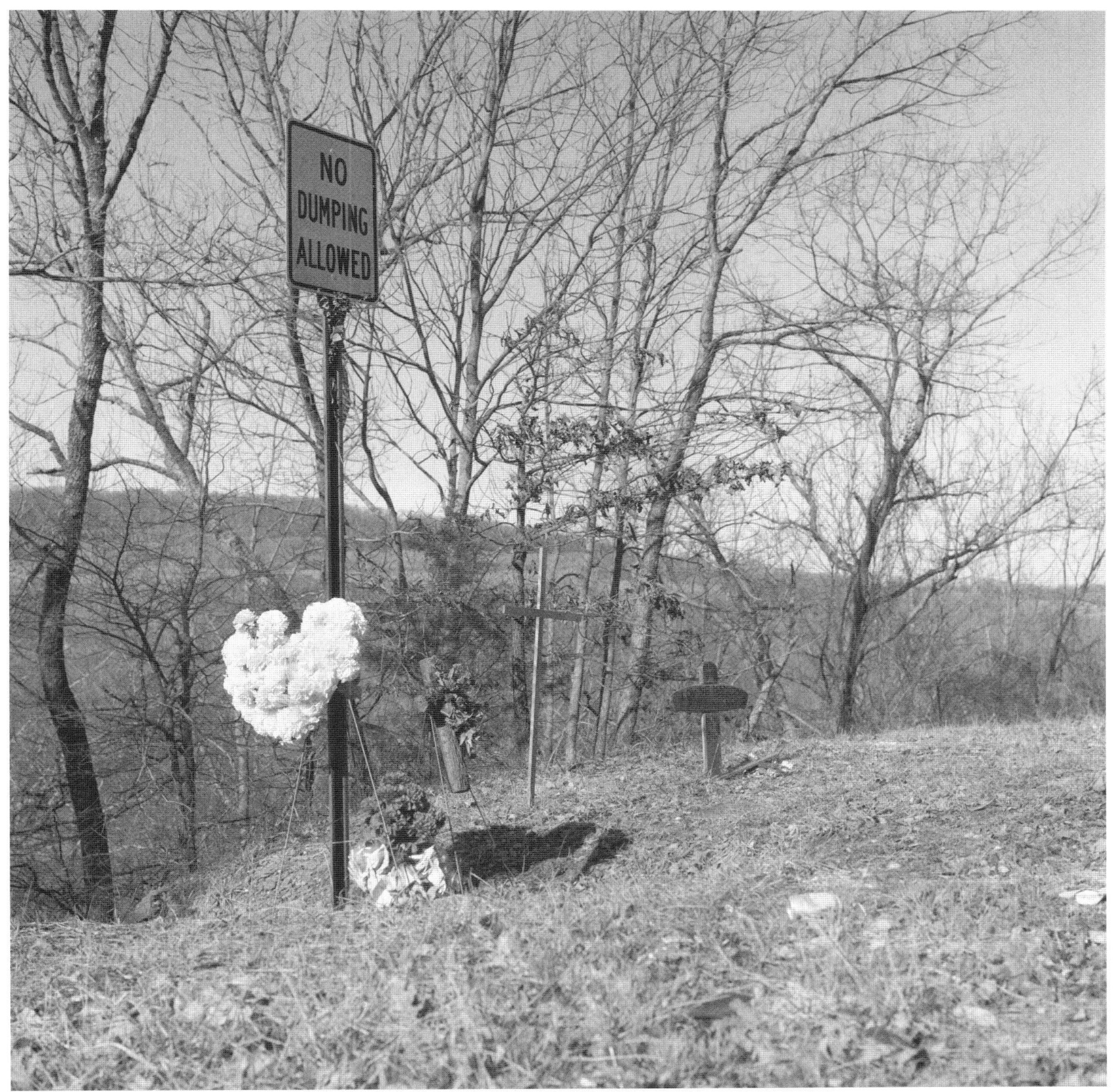

NO
DUMPING
ALLOWED

41

❖ Real estate office

Highway 412, near Huntsville, Arkansas, 2009

The economic downturn hits home.

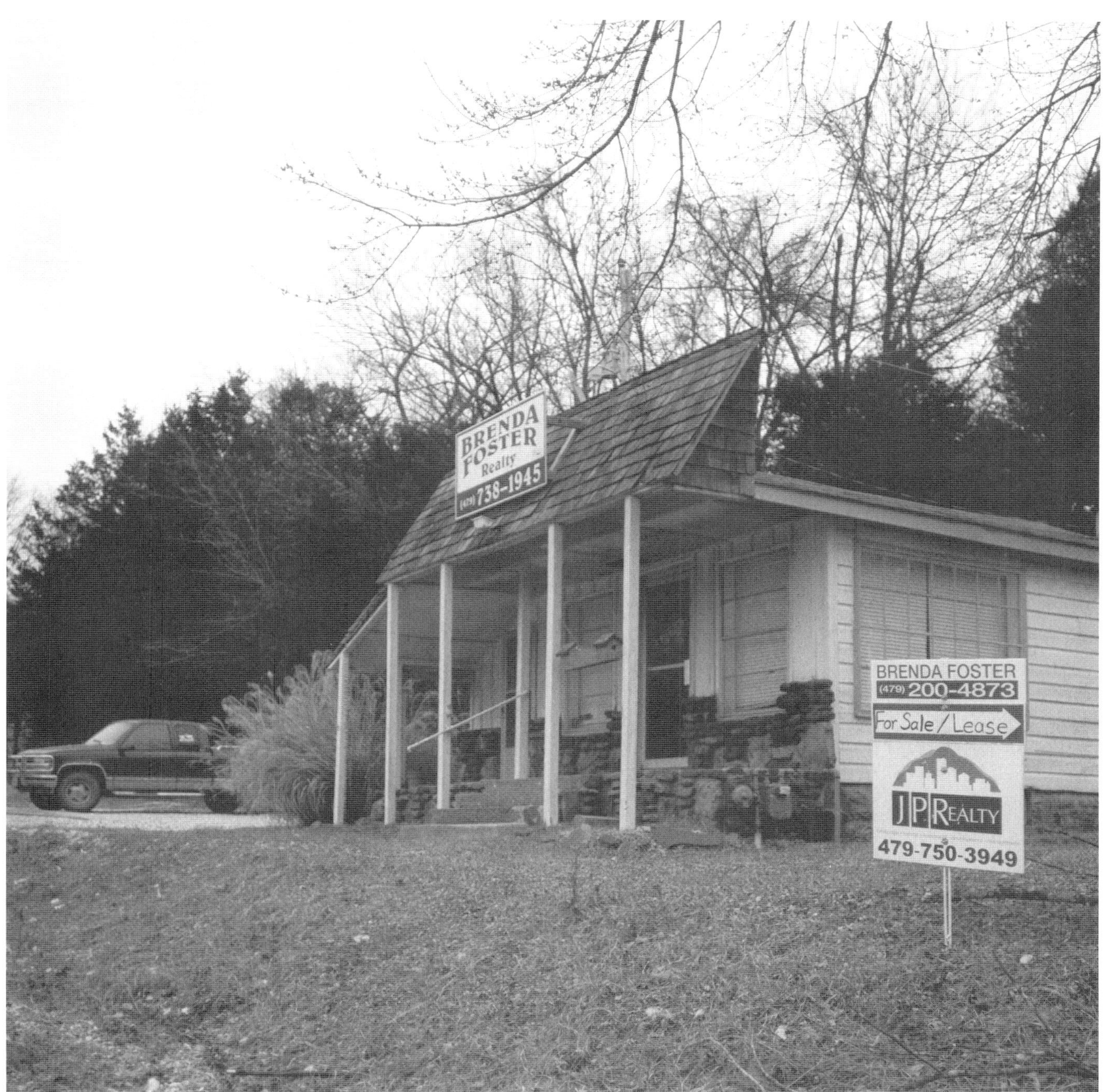

BRENDA
FOSTER
Realty
(479) 738-1945
BRENDA FOSTER
(479) 200-4873
For Sale / Lease
JP REALTY
479-750-3949

Highway 16, Elkins, Arkansas, 2002

No one should be too surprised by the combination of religion and sports, but asking the passerby to think about the brevity of life is fairly heavyweight philosophy-wise. I imagine the president calling a press conference to tell us intelligent life has been found on another planet, but how about them Razorbacks?

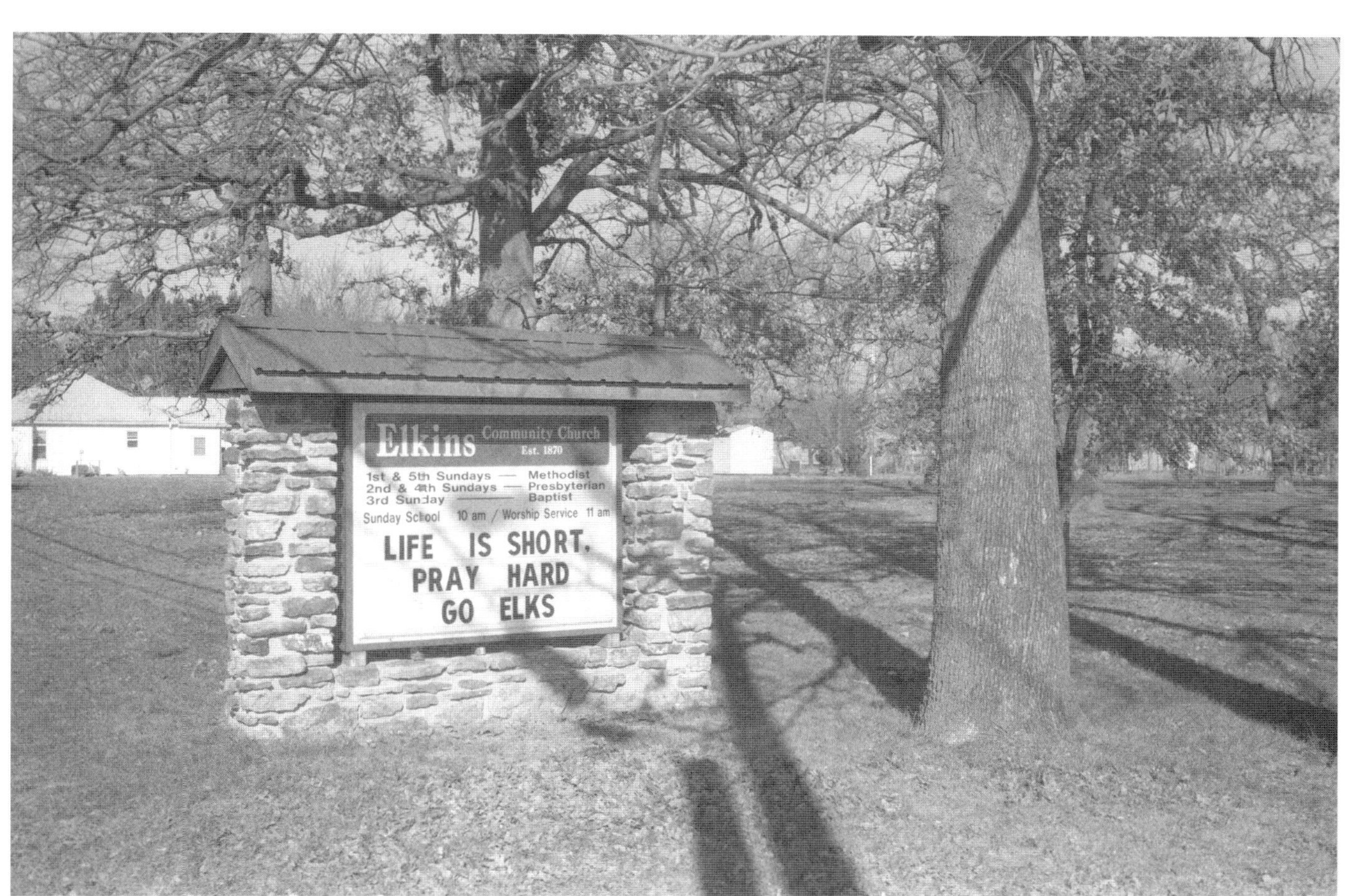
Elkins Community Church
Est. 1870
1st & 5th Sundays — Methodist
2nd & 4th Sundays — Presbyterian
3rd Sunday — Baptist
Sunday School 10 am / Worship Service 11 am
LIFE IS SHORT,
PRAY HARD
GO ELKS

Highway 385, near Burlington, Colorado, 2000

At first, I thought this would be a good example of proof that you're in the middle of nowhere. I love Kansas and southeast Colorado, the flat prairie, big sky beauty of it all, but neighbors are few and far between. However, as I made the mental calculation of what I'd have to do to see Dale Mangus (drive three miles west, two and a half miles northwest, then two miles northeast), it dawned on me that in my own community, I couldn't name all the neighbors in a half-mile radius. So who is living in the middle of nowhere?

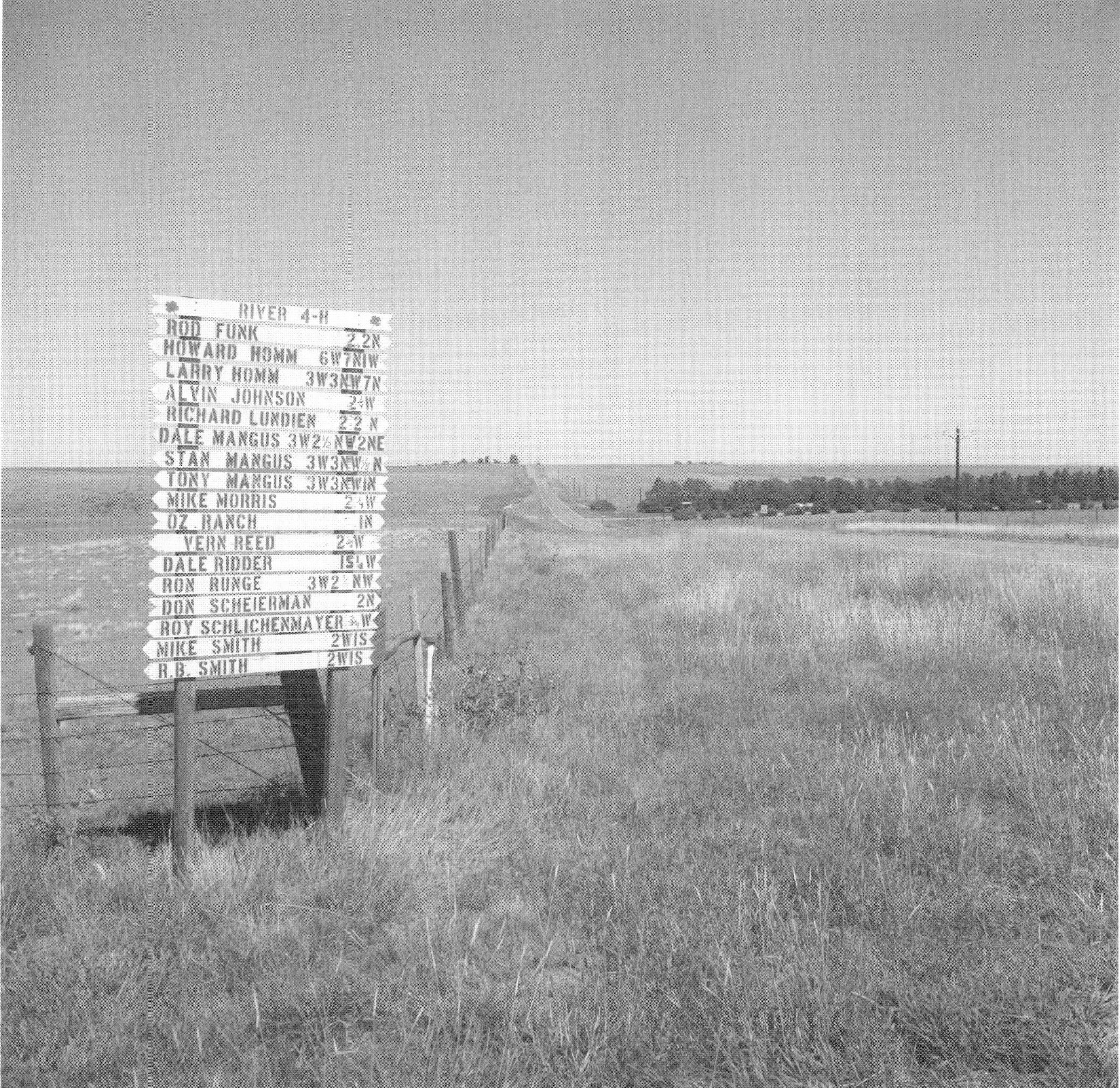

RIVER 4-H
ROD FUNK 2.2N
HOWARD HOMM 6W7N1W
LARRY HOMM 3W3NW7N
ALVIN JOHNSON 2¼W
RICHARD LUNDIEN 2.2 N
DALE MANGUS 3W2½NW2NE
STAN MANGUS 3W3NW½N
TONY MANGUS 3W3NW1N
MIKE MORRIS 2¼W
OZ RANCH 1N
VERN REED 2¼W
DALE RIDDER 1S¼W
RON RUNGE 3W2½NW
DON SCHEIERMAN 2N
ROY SCHLICHENMAYER ¾W
MIKE SMITH 2W1S
R.B. SMITH 2W1S

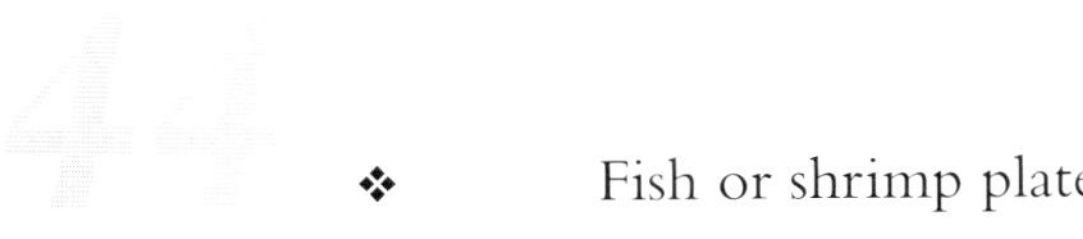

❖   Fish or shrimp plate

Highway 15, Ferriday, Louisiana, 2001

On assignment only a couple of months after the attack of September 11, I watched this expression of support sprout up everywhere, but business just kept on going as usual.

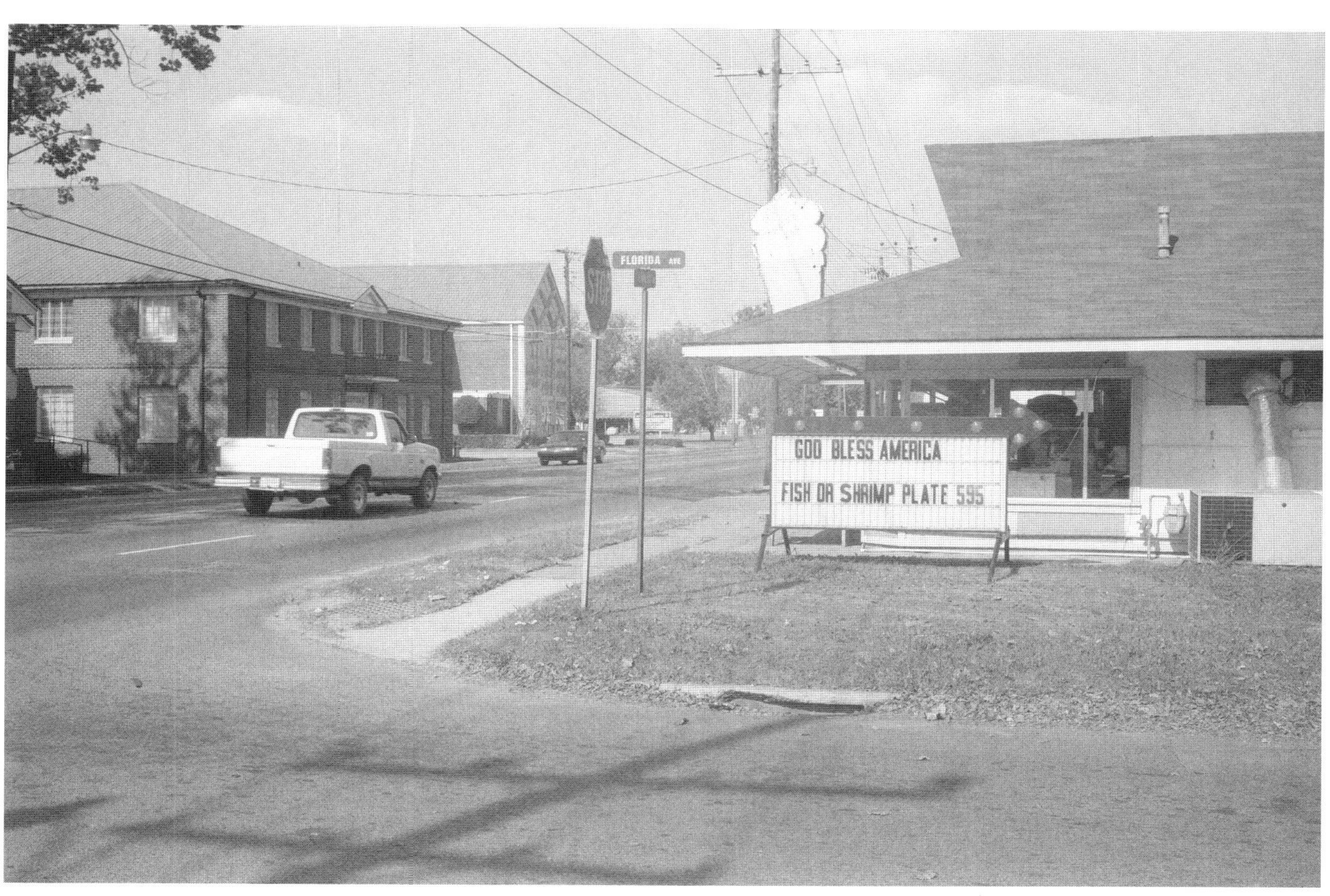

FLORIDA AVE
STOP
GOD BLESS AMERICA
FISH OR SHRIMP PLATE 595

45   ❖   Rex's

Highway 65, south of Branson, Missouri, 2008

It seems like artificial limbs and braces should be made by companies with names like Johnson & Johnson or at research medical centers in major metropolitan areas. There's something very encouraging about knowing that's not necessarily true. Someday I hope to have the courage to stop and ask if I can look over Rex's shoulder as he works.

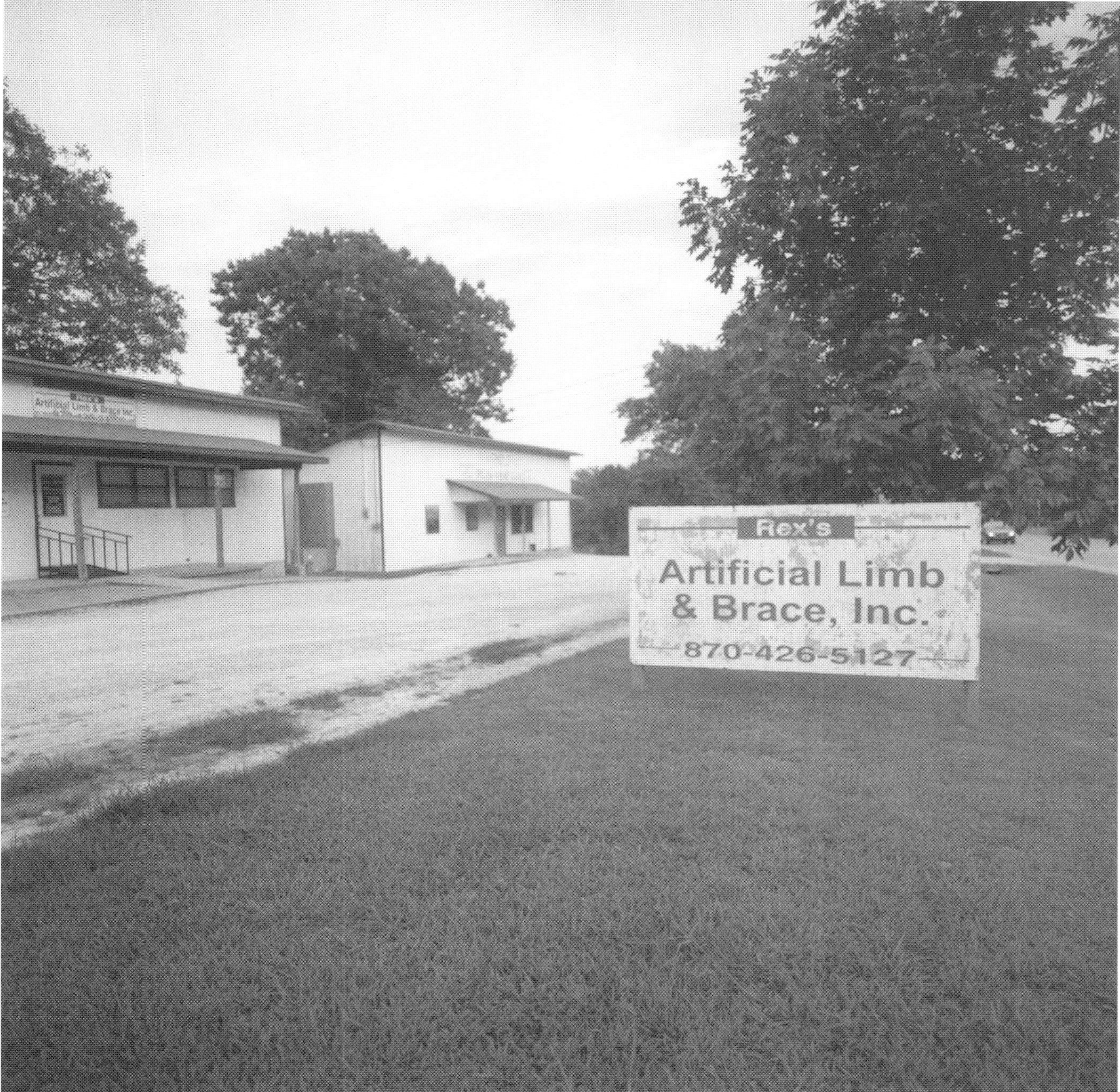

Rex's
Artificial Limb
& Brace, Inc.
870-426-5127

Highway 31, near Quinton, Oklahoma, 2004

I was photographing a gruesome scene west of Quinton—a dozen or more dead coyotes in various states of decay hung on a barb-wire fence along Highway 31, not sure if the landowner was trying to show off to humans or warn off coyotes. A few miles later, looking for a cup of coffee in town, I came across what must be one of the most creative spin control programs ever conceived. The two scenes are connected in someway, I just know it.

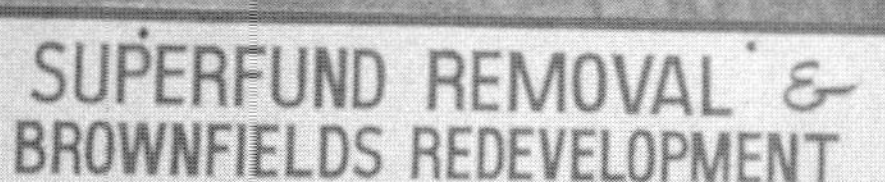

SUPERFUND REMOVAL &
BROWNFIELDS REDEVELOPMENT
Quinton Public Schools & The Quinton Community Thank
US EPA   REGION VI
ODEQ   MARK COLEMAN
QUINTON BOARD OF EDUCATION
ODEd   LELAND TAYLOR
QUINTON CITY COUNCIL
COLIN JANEWAY
JERRY & CAROLYN FOWLER
SUPT. OF SCHOOLS   ART SCHOFIELD
LAURA MILLER
SENATOR   GENE STIPE
STATE REP   LLOYD FIELDS
U.S. CONGRESSMAN   WES WATKINS
City of Quinton
OKLAHOMA
DIGGING TODAY FOR THE FUTURE OF TOMORROW

Highway 27, Tylertown, Mississippi, 2001

The neighbors are quiet, the landlord takes care of the yard, and rent includes all utilities.

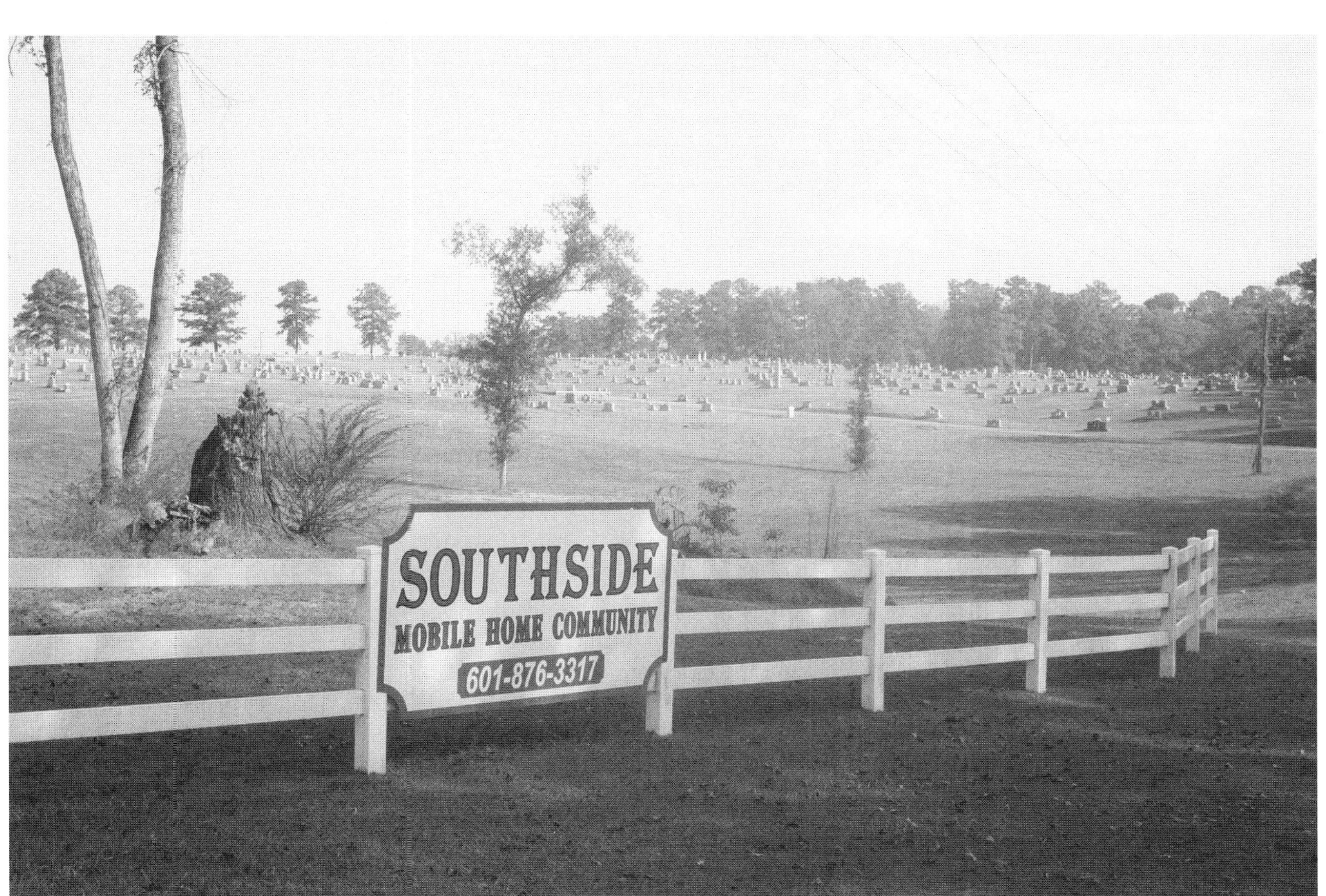
SOUTHSIDE
MOBILE HOME COMMUNITY
601-876-3317

48

❖   View area ahead

Highway 191, near Arches National Park, eastern Utah, 2007

Overheard conversation between a man and woman standing here looking out at one of the most rugged and beautiful landscapes on earth:

    Her: "I don't get it."
    Him: "Me neither, what are we supposed to be viewing?"
    Her: "Let's go."

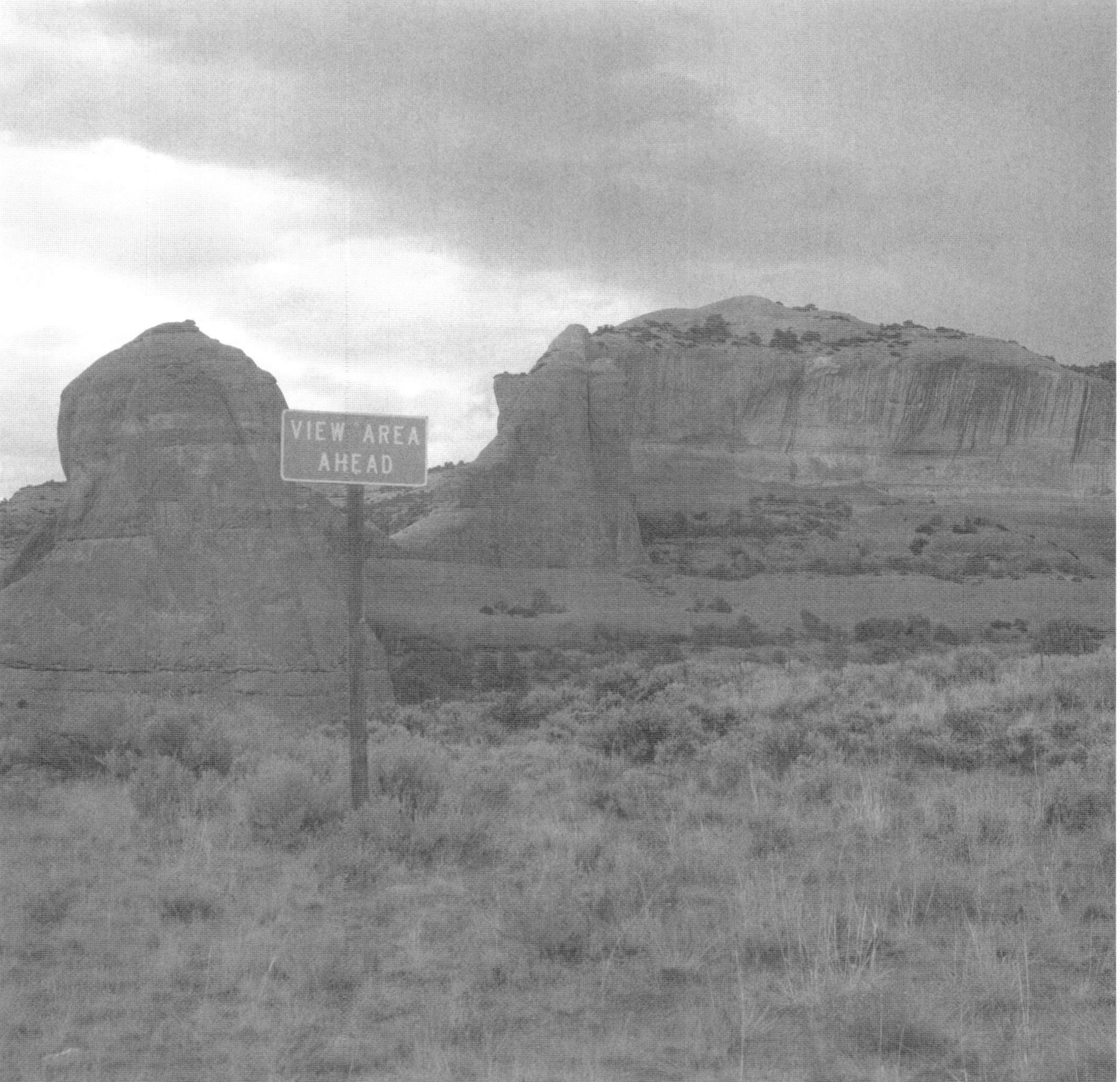

VIEW AREA
AHEAD

49

❖   Stoop & scoop

Highway 22, western Michigan, 2001

Makes a person hesitate to kick off their shoes and run through the sand, don't ya think?

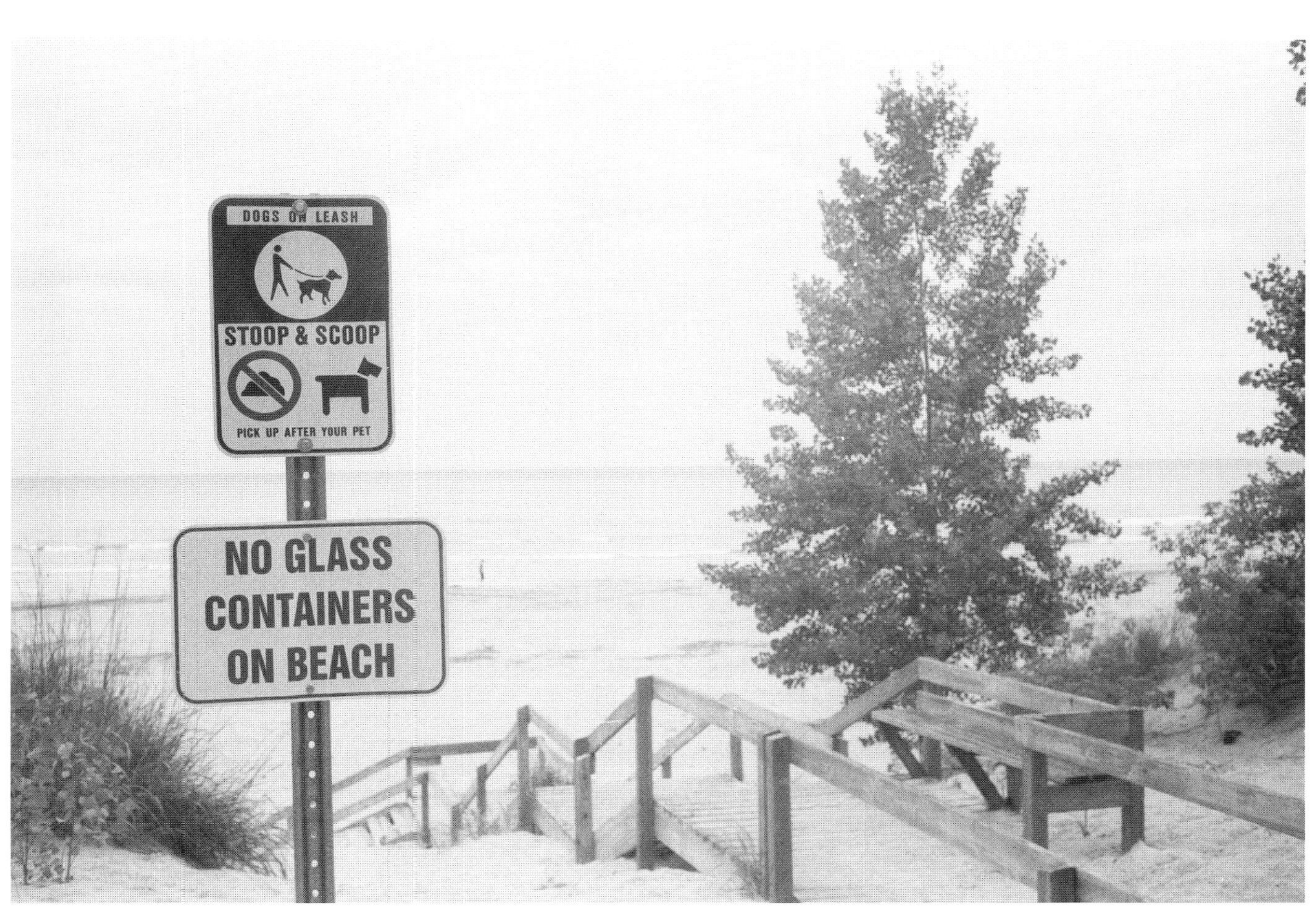

DOGS ON LEASH
STOOP & SCOOP
PICK UP AFTER YOUR PET
NO GLASS
CONTAINERS
ON BEACH

50 ❖ Water valve

Highway 309, Mt. Magazine State Park, near Paris, Arkansas, 2008

The highest point in Arkansas. It's the only place where you can experience a rocky mountain–like phenom-enon: standing at the side of the road in two feet of snow, photographing at a fast shutter speed to compensate for the shivering, then twenty minutes later walking through a warm dry parking lot into the inviting aromas of the Grapevine Restaurant in Paris.

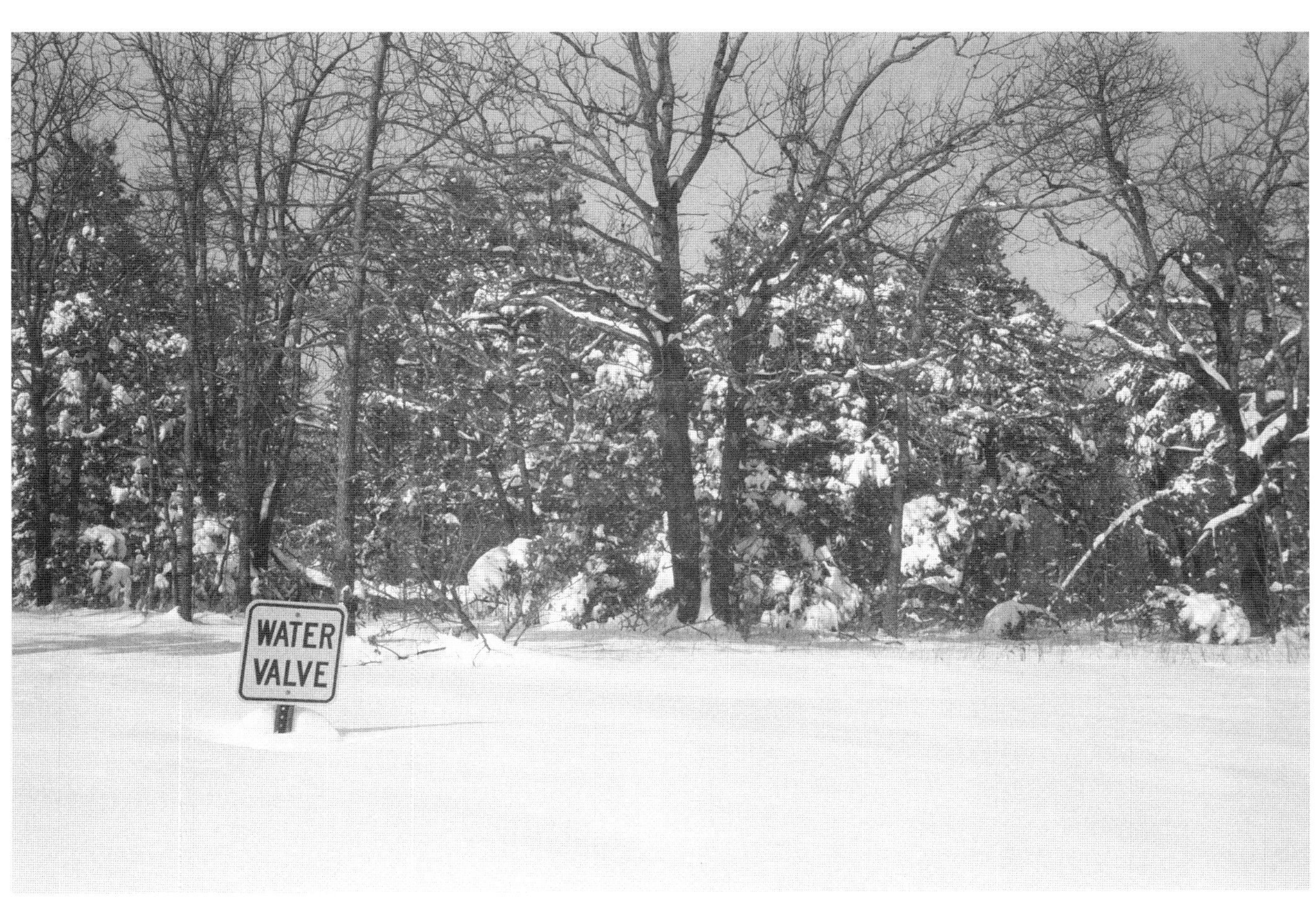

WATER
VALVE

❖   Please cal

Mount Comfort Road, near Tontitown, Arkansas, 2008

There's an expression used by old-timers in Arkansas, usually said after doing something really dumb: "I shouldn't be allowed to vote."

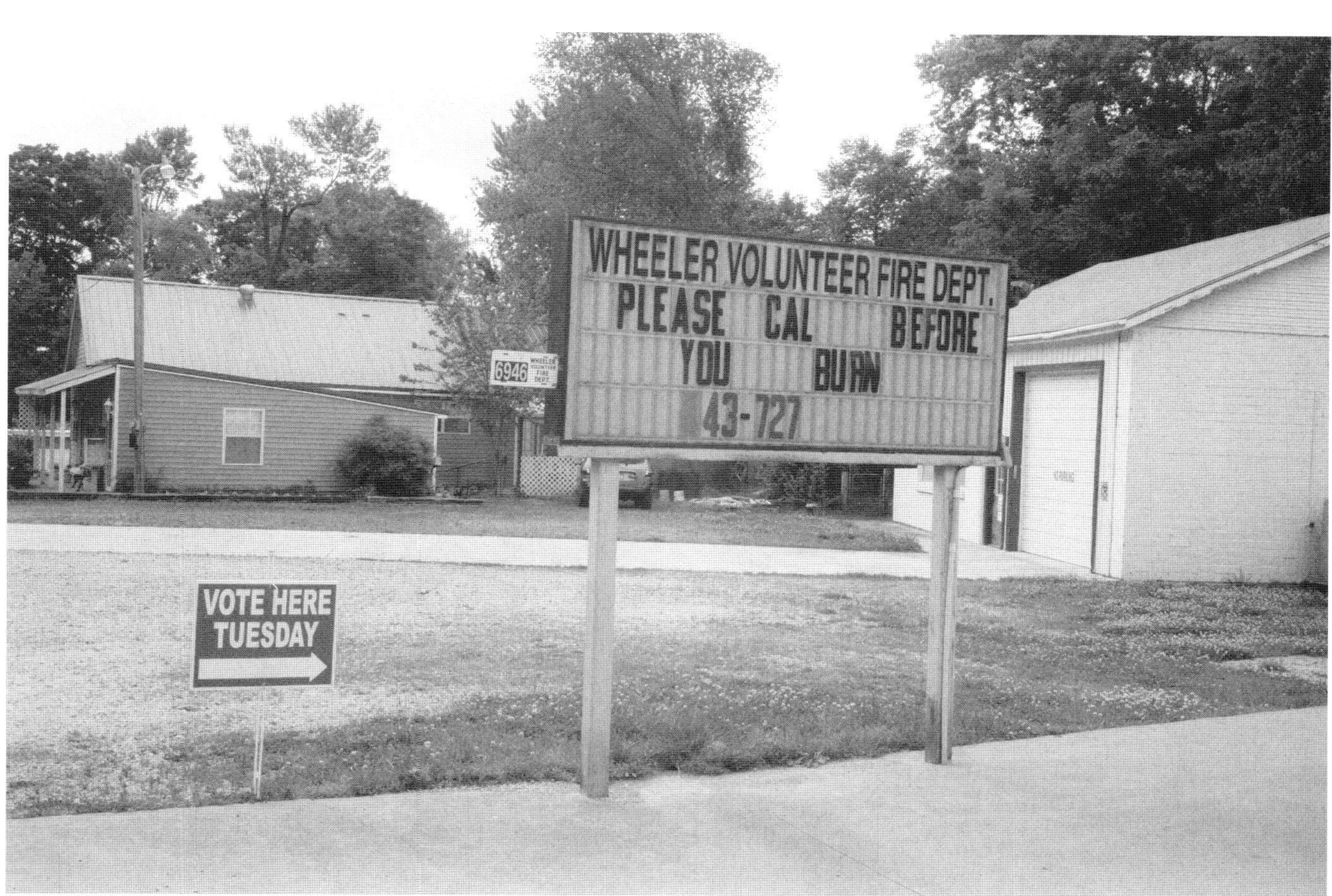
WHEELER VOLUNTEER FIRE DEPT.
PLEASE CAL BEFORE
YOU BURN
43-727
6946
WHEELER
VOLUNTEER
FIRE
DEPT.
VOTE HERE
TUESDAY

Highway 191, southern Utah, 2007

Utah can seem like a country under siege—a wild, beautiful, fragile land where a handful of national park rangers face down an army of off-roaders, miners, drillers, and developers—so when I came around the bend on my way to Arches National Park, I thought, "Damn, too late!"

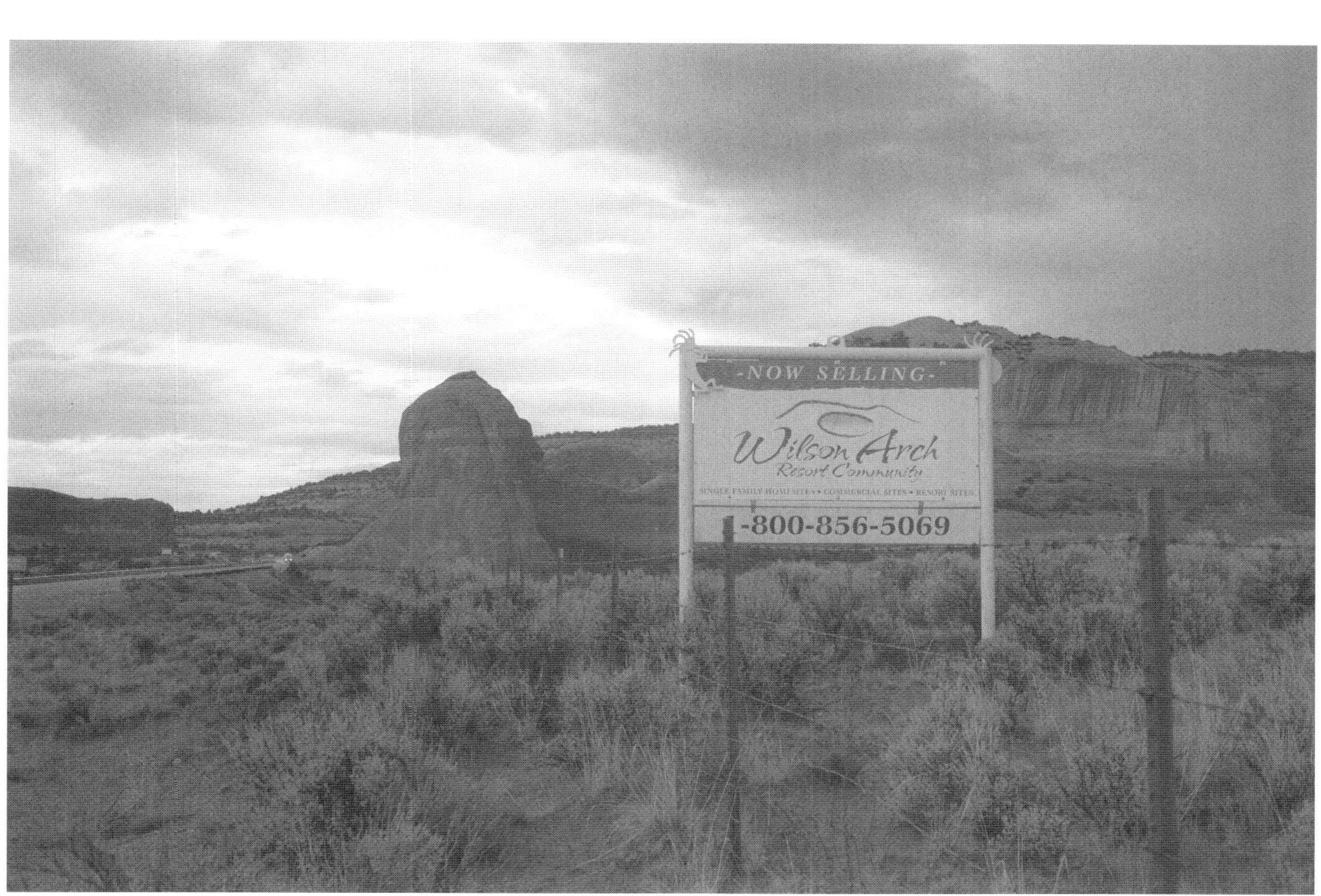

-NOW SELLING-
Wilson Arch
Resort Community
SINGLE FAMILY HOMESITES • COMMERCIAL SITES • RESORT SITES
1-800-856-5069

❖ God lives there

Center Street, Fayetteville, Arkansas, 2007

There's something very wrong here. Is this supposed to be comforting?

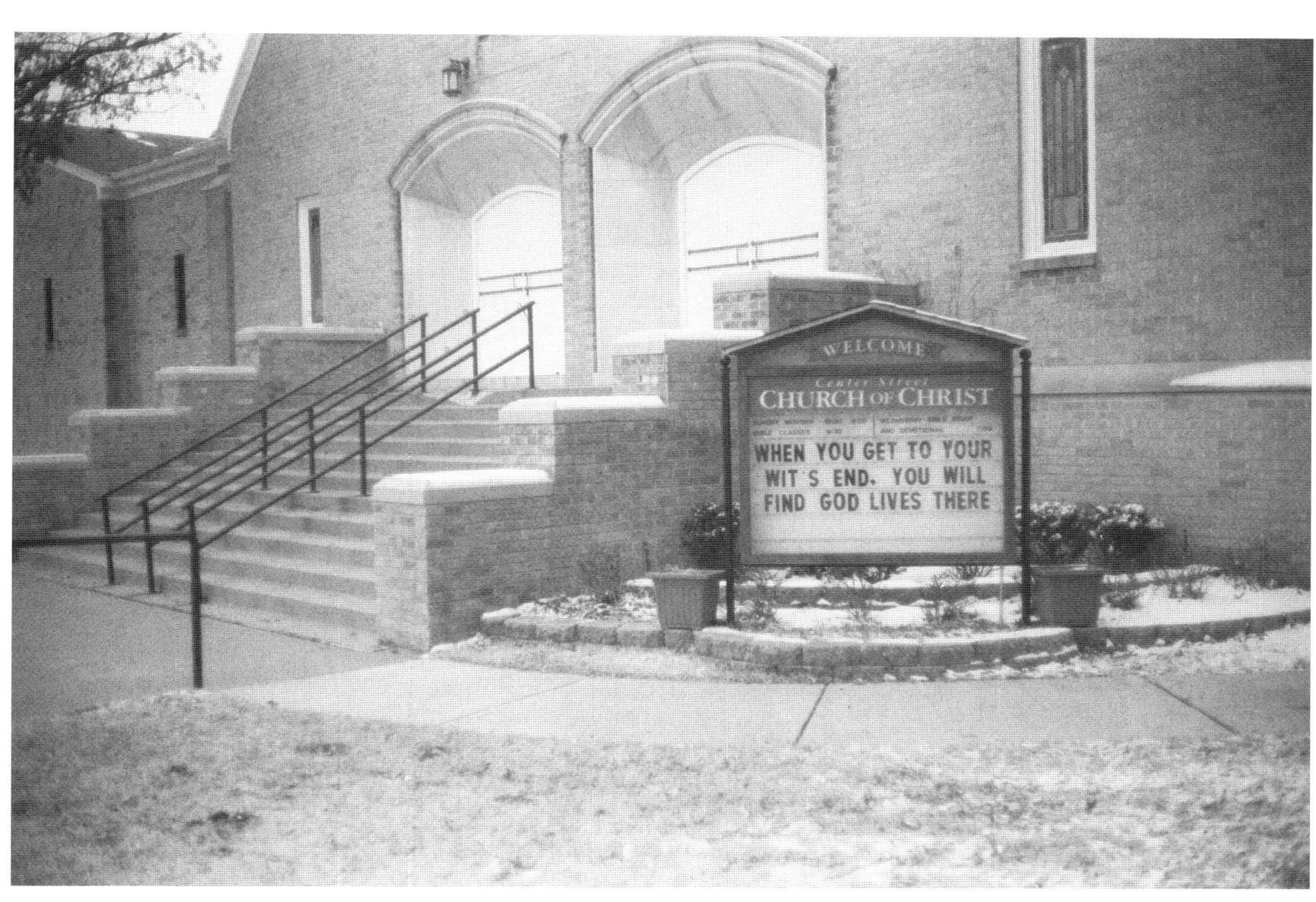

WELCOME
Center Street
CHURCH OF CHRIST
WHEN YOU GET TO YOUR
WIT'S END, YOU WILL
FIND GOD LIVES THERE

Highway 16, near Boston, Arkansas, 2000

Wishful thinking might be the title of this one. After I went through the dance of photographing this out-house—tripod setup, meter reading, composition, darkslide removal, shutter release—I couldn't resist walking over and opening the door. Something about the scale of the building and its placement made me think of a crude trap put there by aliens, hoping to lure human females—or curious photographers.

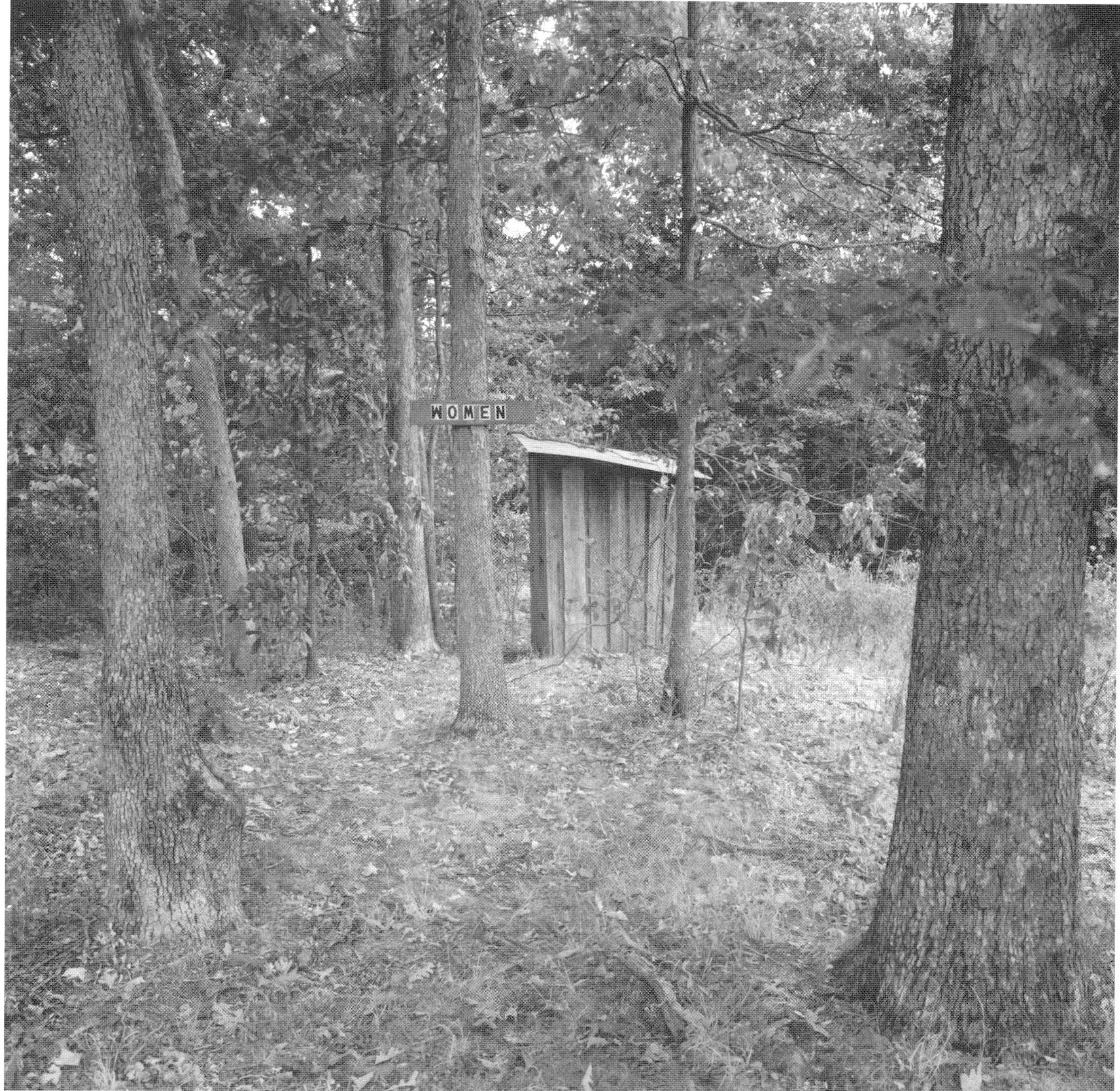

WOMEN

❖   Water may exist

Highway 83, near Scott City, Kansas, 2000

As best I can figure, someone in the county sign shop had gotten a degree in philosophy and was determined to put it to use. Or perhaps, life in Kansas just puts a person into a philosophical frame of mind.

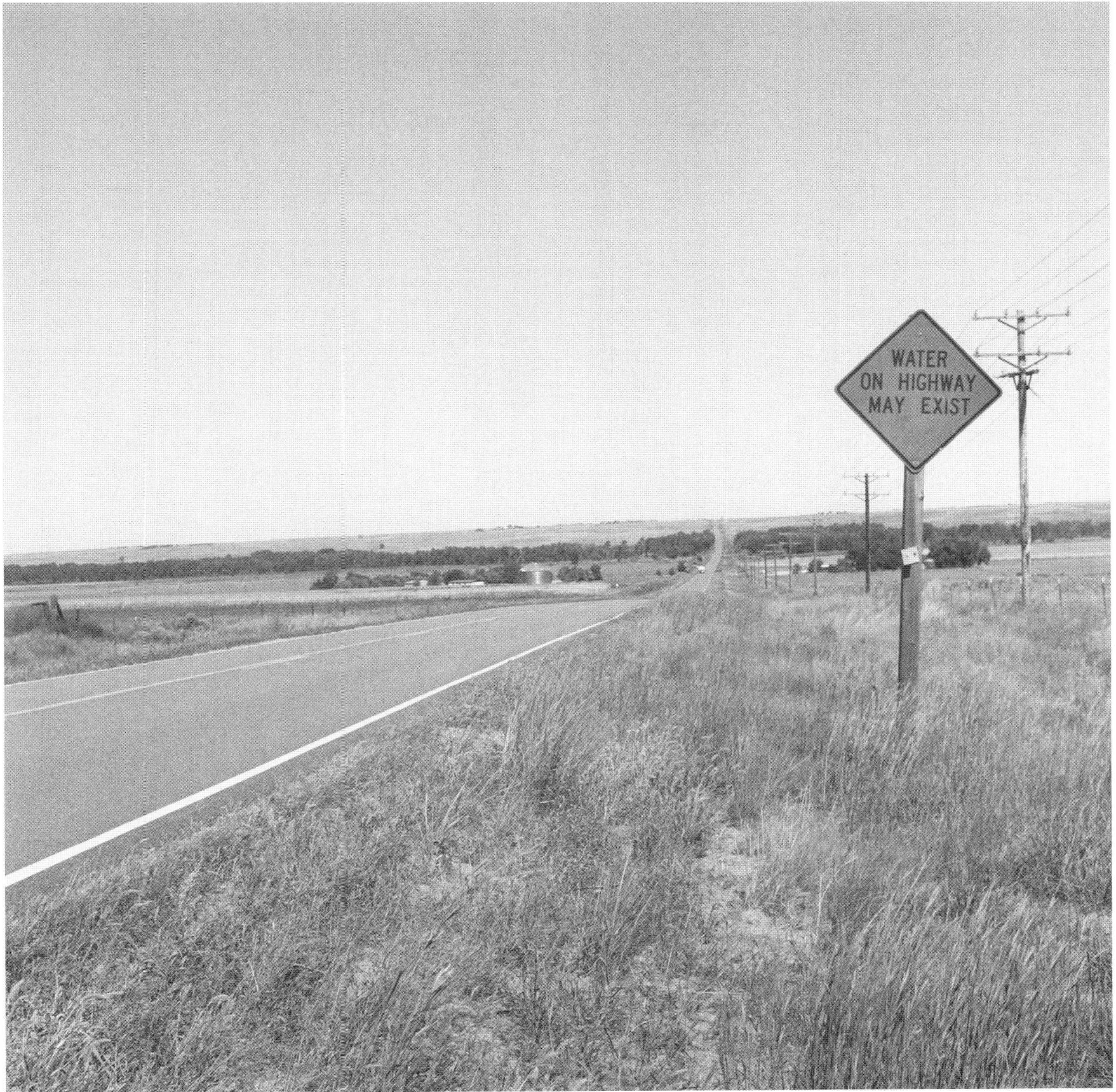

WATER
ON HIGHWAY
MAY EXIST

Highway 5, near Hartville, Missouri, 2009

The couple, sitting on their porch, waved to me, and the scene was so pleasant—early evening, sun dropping— I wanted to pull in, introduce myself, sit, have a coffee and talk politics. What prompted the sign? A local crisis or a general disgust? And what exactly does *think American* mean? I've been a card-carrying American all of my life, so why didn't I feel welcome?

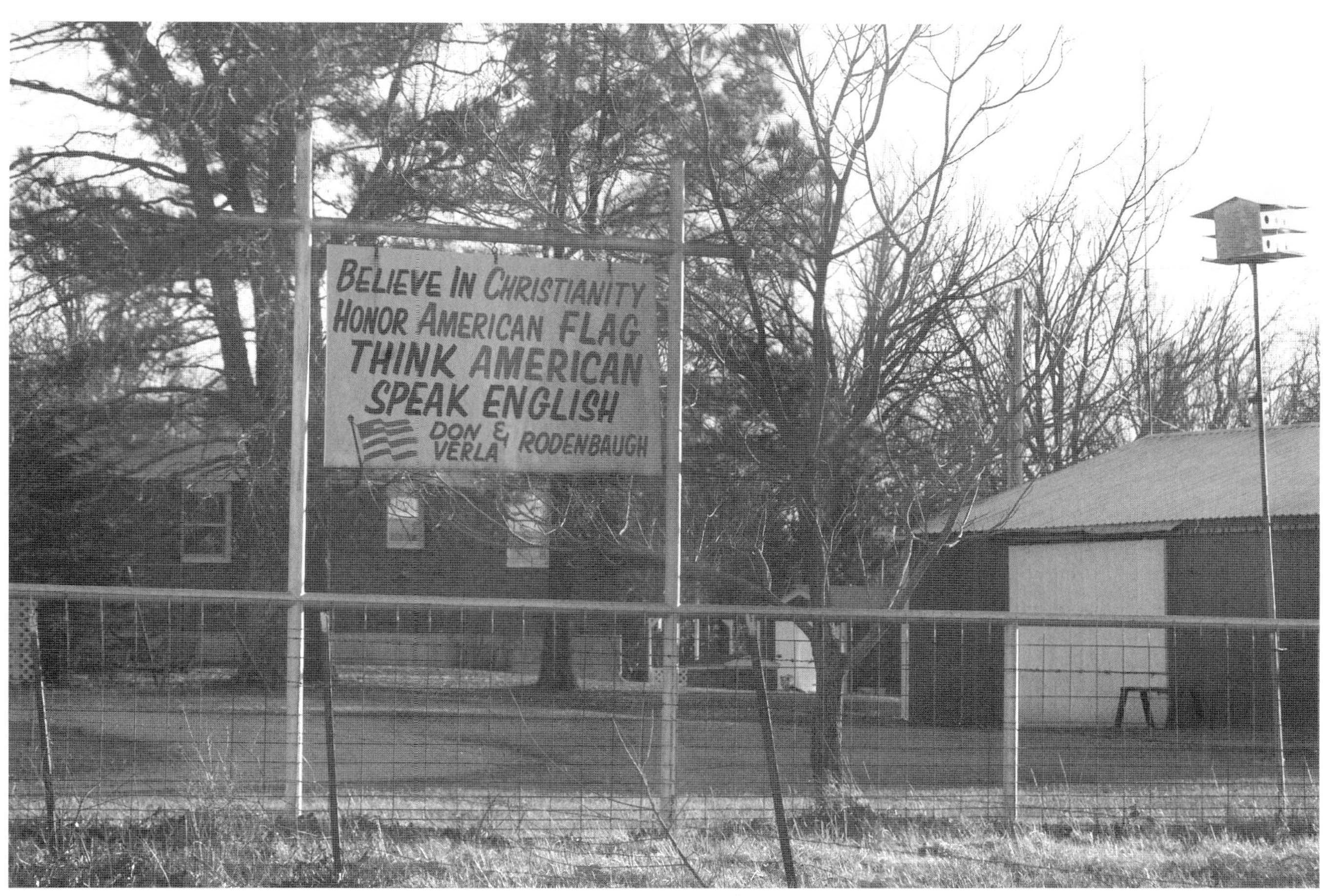
BELIEVE IN CHRISTIANITY
HONOR AMERICAN FLAG
THINK AMERICAN
SPEAK ENGLISH
DON & VERLA RODENBAUGH

Highway 33, near Lagrue, Arkansas, 2008

I never could find the slave cemetery. Maybe they were just being philosophical, or maybe they meant Mississippi.

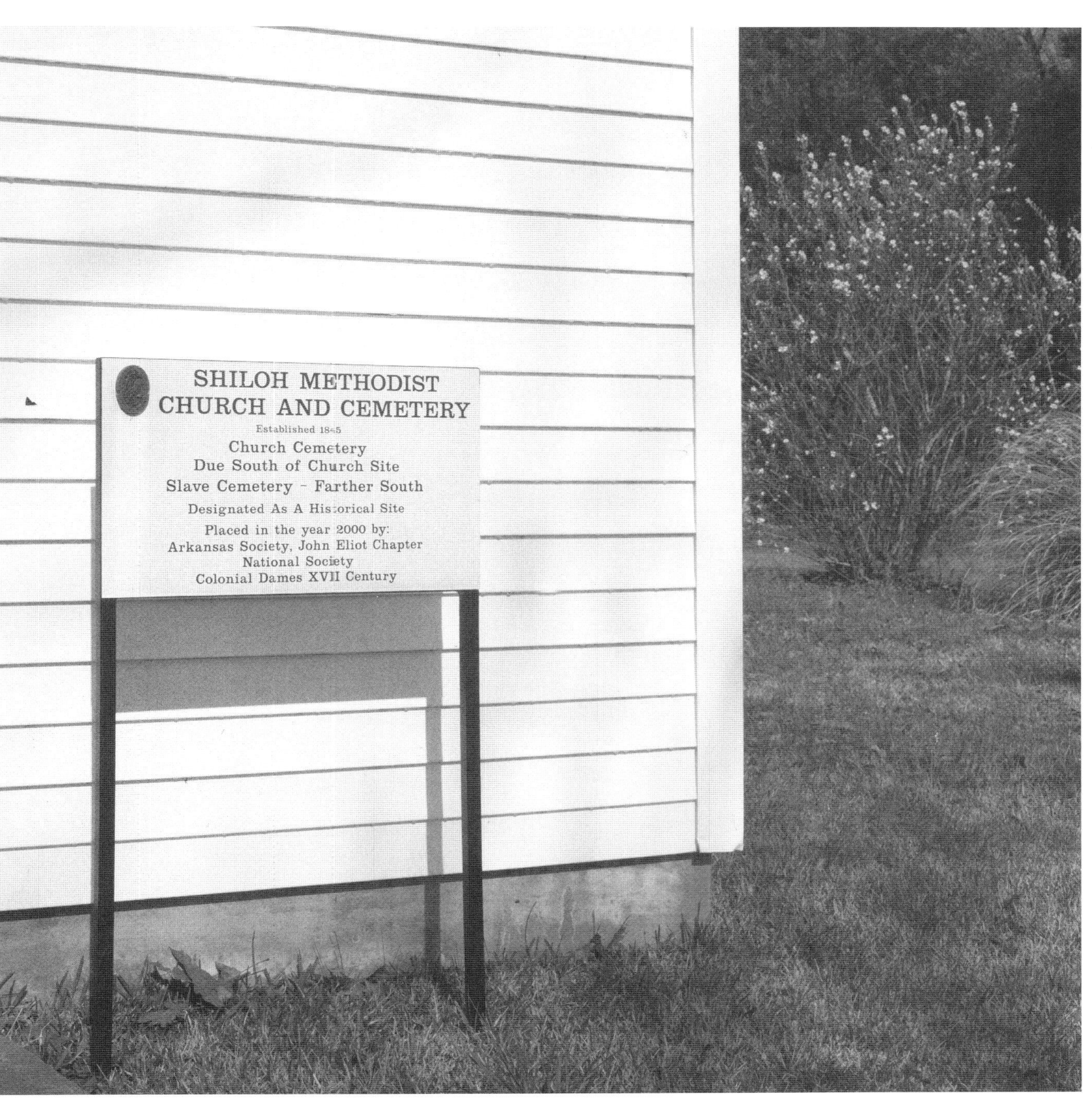
SHILOH METHODIST
CHURCH AND CEMETERY
Established 1845
Church Cemetery
Due South of Church Site
Slave Cemetery - Farther South
Designated As A Historical Site
Placed in the year 2000 by:
Arkansas Society, John Eliot Chapter
National Society
Colonial Dames XVII Century

Highway 24, southern Utah, Capitol Reef National Park, 2007

We all knew it would be something like this. You're walking along, whistling, enjoying life and the incredible beauty that surrounds you, and suddenly there stands a hooded figure holding a scythe. Or perhaps there's only a sign.

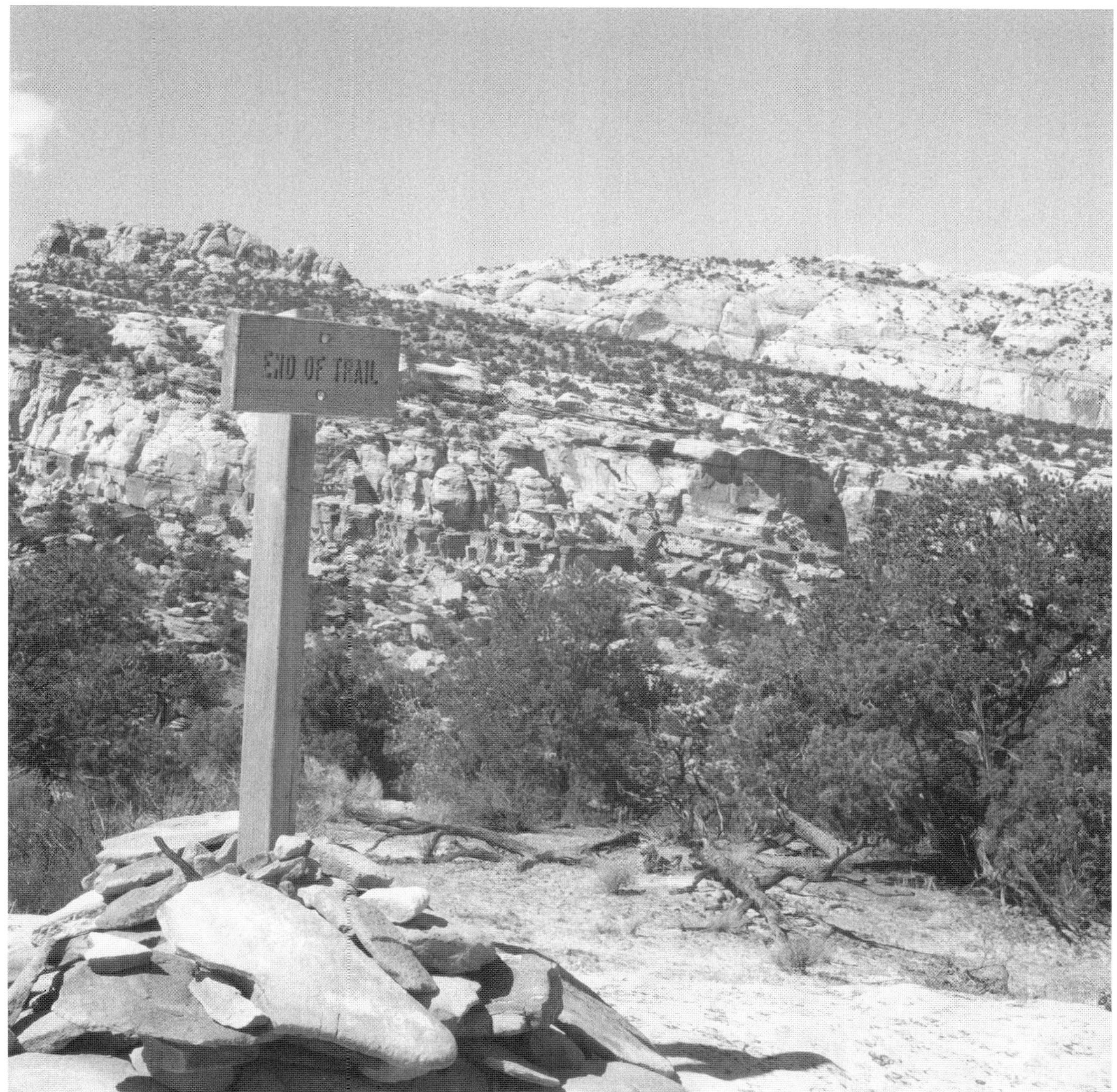
END OF TRAIL

Everyone who dabbles in photography recognizes sooner or later that each camera has its own rhythm, and just like choosing to take the back roads, choice of camera can affect the experience in significant ways. When you find the right one, you tend to stick with it, which is why all of the images in this book were recorded on traditional film using one of two cameras—an ancient Canon 35mm rangefinder, or an equally ancient Hasselblad 6 x 6cm. Both are old friends, workhorses that lend themselves to my rhythm of shooting.

A significant part of that rhythm involves patience and calculation, waiting for the right light, careful exposure, and often a decision to not take it, and return at a better time. As the photographer Russell Cothren once described it: "Learning to be a good photographer means learning when to say 'no.'" Only a few of these photographs followed that advice. When you're passing through, you grab the shot as best you can. Some are enlargements of very small portions of the negative, some were taken through the windshield from the driver's side while in motion, exposures were often guesstimated, some were taken in the worst possible light. In other words they broke my own rules. But I'll have to admit, it was fun.

I have learned the hard way that postponing a photograph is a risky gamble. Perhaps the best example I can offer is the image mentioned in the introduction—*Dwight's Body Shop.* Passing that spot over and over, taking it for granted, waiting for just the right conditions, I arrived on the appointed day only to find it burned to the ground.

Although you haven't asked for it, I offer you my best advice for taking good photographs: *keep your camera handy and your eyes open, and always take the back road.*